APPROACHES TO

FAUNAL ANALYSIS IN THE MIDDLE EAST

edited by

Richard H. Meadow

and

Melinda A. Zeder

Peabody Museum Bulletin 2

Peabody Museum of Archaeology and Ethnology
Harvard University

1978

to BARBARA LAWRENCE
scholar, teacher
and good friend

A current list of all publications available can be obtained
by writing to the Publications Department
Peabody Museum, Harvard University
11 Divinity Avenue, Cambridge, Massachusetts 02138

COPYRIGHT 1978 BY THE PRESIDENT AND FELLOWS OF HARVARD COLLEGE
ISBN 0-87365-951-1
LIBRARY OF CONGRESS CATALOG CARD NUMBER 78-50908
PRINTED IN THE UNITED STATES OF AMERICA

PUBLISHER'S FOREWORD

The long history of publishing advanced work in anthropology for which the Peabody Museum is noted has entered a new phase with a conscious effort being made to broaden the scope of the publications and to report the research of scholars who are not Cambridge-based. The recognition that more traditional formats involve very high list prices for publications that are directed to specialized audiences led to the introduction of the Peabody Museum Bulletin series in 1976. With this format we demonstrate a more economical approach which has been very successful in its initial volume, *A Guide to the Measurement of Animal Bones from Archaeological Sites* by Angela von den Driesch. The present Bulletin, with its more elaborate two-column page format and numerous half-tones, shows the continuing development of this approach.

The two editors, Richard Meadow and Melinda Zeder, are especially commended for their in-depth participation in the publishing process. Virtual membership status on our Publications staff has accrued to them for their editorial efforts as well as for their work on physically putting the book together. Only their total involvement has enabled us to offer this important work. The Museum thanks all the contributors, but especially the two editors for bringing together zooarchaeologists from around the globe who have been doing significant work in the Middle East and for seeing this publication through to completion.

C.C. Lamberg-Karlovsky
Director
April 1978

CONTENTS

PUBLISHER'S FOREWORD by C.C. Lamberg-Karlovsky　*iii*

FOREWORD by Charles A. Reed　*ix*

PREFACE by Richard H. Meadow and
Melinda A. Zeder　*xiii*

PART ONE: *METHODOLOGY*　1

   Øystein Sakala LaBianca
THE LOGISTIC AND STRATEGIC ASPECTS OF FAUNAL
ANALYSIS IN PALESTINE　3

   Barbara Lawrence
ANALYSIS OF UNIDENTIFIABLE BONE FROM ÇAYÖNÜ: AN
EARLY VILLAGE FARMING COMMUNITY　11

   Richard H. Meadow
EFFECTS OF CONTEXT ON THE INTERPRETATION OF FAUNAL
REMAINS: A CASE STUDY　15

PART TWO: *MEASUREMENTS*　23

   Joachim Boessneck and Angela von den Driesch
THE SIGNIFICANCE OF MEASURING ANIMAL BONES FROM
ARCHAEOLOGICAL SITES　25

   Hans-Peter Uerpmann
METRICAL ANALYSIS OF FAUNAL REMAINS FROM THE MIDDLE
EAST　41

PART THREE: *DOMESTICATION AND ENVIRONMENT*　47

   Juliet Clutton-Brock
BONES FOR THE ZOOLOGIST　49

   Pierre Ducos
"DOMESTICATION" DEFINED AND METHODOLOGICAL
APPROACHES TO ITS RECOGNITION IN FAUNAL
ASSEMBLAGES　53

   Sándor Bökönyi
ENVIRONMENTAL AND CULTURAL DIFFERENCES AS REFLECTED
IN THE ANIMAL BONE SAMPLES FROM FIVE EARLY
NEOLITHIC SITES IN SOUTHWEST ASIA　57

   Richard W. Redding
RODENTS AND THE ARCHAEOLOGICAL PALEOENVIRONMENT:
CONSIDERATIONS, PROBLEMS, AND THE FUTURE　63

   Melinda A. Zeder
DIFFERENTIATION BETWEEN THE BONES OF CAPRINES FROM
DIFFERENT ECOSYSTEMS IN IRAN BY THE ANALYSIS OF
OSTEOLOGICAL MICROSTRUCTURE AND CHEMICAL
COMPOSITION　69

PART FOUR: *PAPERS ON FAUNAL REMAINS FROM
SHAHR-I SOKHTA*　85

   Lucia Caloi, Bruno Compagnoni,
     and Maurizio Tosi
PRELIMINARY REMARKS　87

   Bruno Compagnoni and Maurizio Tosi
THE CAMEL: ITS DISTRIBUTION AND STATE OF
DOMESTICATION IN THE MIDDLE EAST DURING THE THIRD
MILLENNIUM B.C. IN LIGHT OF FINDS FROM
SHAHR-I SOKHTA　91

   Bruno Compagnoni
THE BONE REMAINS OF EQUUS HEMIONUS FROM SHAHR-I
SOKHTA　105

   Bruno Compagnoni
THE BONE REMAINS OF GAZELLA SUBGUTTUROSA FROM
SHAHR-I SOKHTA　119

   Lucia Caloi
THE BONE REMAINS OF SMALL WILD CARNIVORES FROM
SHAHR-I SOKHTA　129

PART FIVE: *CODING SYSTEMS*　133

   Richard W. Redding, Melinda A. Zeder, and
   John McArdle
"BONESORT II" -- A SYSTEM FOR THE COMPUTER
PROCESSING OF IDENTIFIABLE FAUNAL MATERIAL　135

   Hans-Peter Uerpmann
THE "KNOCOD" SYSTEM FOR PROCESSING DATA ON ANIMAL
BONES FROM ARCHAEOLOGICAL SITES　149

   Richard H. Meadow
"BONECODE" -- A SYSTEM OF NUMERICAL CODING FOR
FAUNAL DATA FROM MIDDLE EASTERN SITES　169

FIGURES

PART TWO

Boessneck and von den Driesch: THE SIGNIFICANCE OF MEASURING ANIMAL BONES
1. Decrease over time in the size of red deer from the finds of Magulla Pevkakia, Thessaly — 27
2. Size comparison between prehistoric red deer bones from the Iberian Peninsula and from Central Europe — 28
3. Sexual dimorphism in metacarpals and metatarsals of red deer — 29
4. Size differences between aurochs and Copper Age domestic cattle in Portugal and sexual dimorphism in the astragalus — 30
5. Sexual dimorphism in the metacarpus of prehistoric cattle from Zambujal, Portugal — 30
6. Cattle metacarpals from Central Europe and Rumania, Roman Period from Roman-occupied territory — 31
7. Cattle metacarpals from Central and Northern Europe, Roman Period but from outside of Roman-occupied territory — 31
8. Variation in the length of the molar row in mandibles of domestic cattle from Europe — 32
9. Variation in the greatest length of metatarsus of domestic cattle from Europe — 32
10. Comparison of greatest length of metacarpus in cattle from an urban and a rural settlement in Schleswig-Holstein from the early Middle Ages — 33
11. Separation of ptarmigan species *Lagopus lagopus* and *Lagopus mutus* on the basis of the tarsometatarsus — 34
12. Variation in size and proportions in the carpometacarpus of smaller galliformes of North and Central Europe — 35

Uerpmann: METRICAL ANALYSIS OF FAUNAL REMAINS
1. Size variability of southwest asiatic sheep represented by measurements of the astragalus — 43

PART THREE

Bökönyi: ENVIRONMENTAL AND CULTURAL DIFFERENCES
1. Graphic representations of faunal proportions from five sites in Southwest Asia — 58

Zeder: DIFFERENTIATION BETWEEN THE BONES OF CAPRINES
1. Map of collecting localities of the lowland sedentary and wild sheep and the routes of the transhumant sheep used in the study — 71
2. Longitudinal thin sections of the distal humeri of sheep as viewed through the polarizing microscope — 74
3. Longitudinal cut from the distal humerus of a sheep as viewed with a scanning electron microscope — 74
4. Cross section from the distal humerus of a sheep as viewed through a polarizing microscope — 75
5. Means and standard deviations of the bone/space ratios for each ecological group — 76
6. Means and standard deviations of Calcium, Magnesium, and Zinc content of the bones for each ecological group — 79

PART FOUR

Caloi, Compagnoni, and Tosi: PRELIMINARY REMARKS
1. Approximate proportions of faunal remains recovered using different techniques — 89

Compagnoni and Tosi: THE CAMEL
1. Bones of *Camelus* sp. from Shahr-i Sokhta — 93
2. Camel dung from Shahr-i Sokhta and comparative material — 94
3. Clay animal figurine from Shahr-i Sokhta — 94
4. Textile from Shahr-i Sokhta — 97
5. Limestone orthostats from Umm an-Nar — 97
6. Map of the reconstructed areas of distribution of *Camelus bactrianus* and *Camelus dromedarius* — 99

Compagnoni: THE BONE REMAINS OF EQUUS HEMIONUS
1. Occipital measurements of *Equus hemionus* — 108
2. Bones of *Equus hemionus* from Shahr-i Sokhta (I) — 110
3. Bones of *Equus hemionus* from Shahr-i Sokhta (II) — 112
4. Measurements of first phalanges of onager, donkey, and horse — 113

Compagnoni: THE BONE REMAINS OF GAZELLA SUBGUTTUROSA
1. Bones of *Gazella subgutturosa* from Shahr-i Sokhta (I) — 121
2. Bones of *Gazella subgutturosa* from Shahr-i Sokhta (II) — 122
3. Possible representations of *Gazella* on painted Buff Ware sherds — 125

Caloi: THE BONE REMAINS OF SMALL WILD CARNIVORES
1. Bones of small wild carnivores from Shahr-i Sokhta — 130

PART FIVE

Meadow: "BONECODE"
1. Input and output codes for mandibular tooth wear of sheep and goats — 183

TABLES

PART ONE

Meadow: EFFECTS OF CONTEXT ON THE INTERPRETATION OF FAUNAL REMAINS
1. Counts of identified *Ovis/Capra* and *Sus* bone finds from Hajji Firuz — 17
2. Summary counts of bone finds in Table 1 — 18
3. Percentages of taxa within context type, calculated from Table 2 — 18
4. Percentages within taxon by context type, calculated from Table 2 — 18
5. Summary statistics calculated from Table 2 — 18
6. Bone weights of mammal bones from Hajji Firuz — 20

PART TWO

Boessneck and von den Driesch: THE SIGNIFICANCE OF MEASURING ANIMAL BONES
1. Variation in withers height in populations of pre- and early historic domestic animals — 26
2. Measurements of the first phalanx of red deer from Magula Pevkakia, Thessaly — 27

PART THREE

Ducos: "DOMESTICATION" DEFINED
1. Age-group distributions for Near Eastern settlements — 55

Bökönyi: ENVIRONMENTAL AND CULTURAL DIFFERENCES
1. The percentages of domestic animals and of the major groups of wild animals — 58
2. Faunal lists (presence/absence) — 60

Redding: RODENTS AND THE ARCHAEOLOGICAL PALEOENVIRONMENT
1. Counts by phase and zone of mandibles and maxillae of *Tatera indica* and *Meriones crassus* from Tepe Ali Kosh — 66

Zeder: DIFFERENTIATION BETWEEN THE BONES OF CAPRINES
1. Mean and standard deviation of bone/space ratio and shaft depth x shaft width, shaft depth/shaft width, and thickness of shaft by specimen — 75
2. Mean and standard deviation of the figures in Table 1 by group — 76
3. Significant student's *t*-tests and corresponding ratios between sample variances for two strata of bone/space ratio — 76
4. Correlation coefficients of the figures in Table 1 for wild specimens — 77
5. Mean and standard deviation of Ca, Mg, and Zn by specimen — 78
6. Mean and standard deviation of Ca, Mg, and Zn by group — 78
7. Significant student's *t*-tests and corresponding ratios between sample variances for two strata of mineral content (Ca and Mg) — 78

PART FOUR

Caloi, Compagnoni, and Tosi: PRELIMINARY REMARKS
1. Shahr-i Sokhta, the chronological-stratigraphic sequence — 89
2. Shahr-i Sokhta, chronological distribution of the less well-represented macromammal finds — 89

Compagnoni and Tosi: THE CAMEL
1. Bones of *Camelus* sp. from Shahr-i Sokhta — 92
2. Zoomorphic figurines from Shahr-i Sokhta — 96
3. Evidence for the camel in the third millennium in eastern Iran and southern Turkmenia — 98
4. Evolution of Bedouin society, after Dostal — 100
5. Measurements of camel bones from Shahr-i Sokhta and comparative collections — 101

Compagnoni: THE BONE REMAINS OF *EQUUS HEMIONUS*
1. Bone remains of *Equus hemionus* from Shahr-i Sokhta — 106
2. List of equid specimens used for comparison — 107
3. Skull measurements for Shahr-i Sokhta and modern half-asses — 109
4. Post-cranial measurements for Shahr-i Sokhta and modern half-asses — 114

Compagnoni: THE BONE REMAINS OF *GAZELLA SUBGUTTUROSA*
1. List of gazelle specimens used for comparison — 120
2. Bone remains of *Gazella subgutturosa* from Shahr-i Sokhta — 124
3. Skull measurements of Shahr-i Sokhta gazelles — 125
4. Post-cranial measurements of Shahr-i Sokhta and other gazelles — 126

Caloi: THE BONE REMAINS OF SMALL WILD CARNIVORES
1. Bone remains of wild carnivores from Shahr-i Sokhta — 130
2. *Lutra lutra* measurements — 131
3. *Vulpes vulpes* measurements — 131
4. *Felis* cf. *catus* measurements — 131

PART FIVE

Uerpmann: THE "KNOCOD" SYSTEM
1. Coding for a complete dog skull without teeth — 151
2. KNOCOD -- Order of entry for measurements — 160

Meadow: "BONECODE"
1. BONECODE input format — 170
2. BONECODE input and output codes — 173

FOREWORD

The book you hold in your hand is the end-product of a bright idea which came in 1974 to two students, one graduate, one undergraduate. These same two, Richard Meadow and Melinda Zeder, are now the editors of the finished book born from that idea. Between the idea and the book stands a mountain of work, now successfully accomplished. They organized an international conference on problems of zooarchaeology in the Middle East, they got people to write papers specifically for the conference, they (miracle of miracles!) ferreted out the money to bring several people from Europe to the conference and get them home again, and they got the Society for American Archaeology to sponsor the conference as part of its annual meeting in 1975. They put all of this together into one of the most interesting conferences I have ever attended; then they took the papers, edited them, and convinced the Peabody Museum at Harvard University to publish the results as a book.

The papers here, taken together, show with what increasing strides the science of archaeological faunal analysis has been changing and progressing since only 1963 when, in the first edition of *SCIENCE IN ARCHAEOLOGY*, I offered what now appears as a simplistic approach to enlarging the role of the recovery and study of the skeletal remains of non-human vertebrates from archaeological sites.

The faunal analyst must now be more, much more, than a counter and measurer and identifier of bones. One must be an ecologist, but more; one must be a paleo-environmentalist who aids the palynologist and geomorphologist in the reconstruction of past environments. One must be a student of animal behavior, to reconstruct the living patterns of wild populations and, for some of those, of their domestic descendants. One must be an anatomist, not only to identify bits of broken bones as to species and place in the skeleton, but if possible as to sex and age, and to recognize and understand evidences of pathology. As an anatomist one must be more, however; one must be an evolutionary morphologist, perceiving different and changing patterns of growth and size as correlated with climates, soils, foods, conditions of freedom or servitude, and numbers of generations of such conditions of servitude.

One must also be an archaeologist, to reconstruct settlement patterns, hunting patterns, herding patterns, the flow of parts of butchered carcasses through a village or town, to put the data not only into biological terms, but also into human and humanistic terms, to project one's own thoughts into the past and into the minds of people dead for many millennia, people whose cultural patterns are too often only dimly perceived.

In another direction, one must increasingly become a statistician and a user (hopefully a master) of the computer, to be able to handle quickly, accurately, and logically the tremendous number and variety of facts which emerge from the study of the faunal remains being excavated. One also has to be quite a linguist; faunal studies used to be practiced almost entirely by Central Europeans and published in German, but now everyone is doing them, and publishing the results in numerous languages.

We have achieved much in the last few decades, and, since our conference of 1975, I have met new graduate students who are achieving more. However, problems both major and minor remain to puzzle us. A fundamental one, for instance, is a knowledge of demographic patterns, as expressed in life-tables, of the populations hunted and/or kept by different prehistoric humans as compared to the age-groups killed by the hunters or the butchers of domesticated stock. Without this knowledge, which for the most part we do not yet have, we cannot reconstruct the demography of the living populations of animals whose bones we dig from archaeological sites.

Another difficult problem, whose solution depends in part on the answer to the one presented in the paragraph above, is that of the domestication of animals - when and how, and perhaps, why? Many suggestions have been proposed as to how and why humans domesticated other animals, thus entering relatively recently into a symbiotic relationship unknown before. Some think it was a long, slow, and gradual process of habituation with its beginnings well back into the Palaeolithic; some have thought such domestication, at least for the hoofed food-animals, could not have begun until after humans grew crop-plants and had some surpluses to feed the animals, under which conditions domestication then happened rapidly. Other suggestions fall between these extremes in time and duration of the process.

Before the major problem of the origin of domestication can be attacked, the archaeozoologists have to discover valid and multiple criteria for that state we call domestication (a problem of definition discussed in this volume by Pierre Ducos); domestication is not and probably never was a simple and standard situation but varied widely (as true also today) from close confinement to feral near-wildness. One possibility of such differentiation of domestic from wild suggested by several investigators, with varying results, is that of micromorphological and chemical differences in the bones of animals which lived under different conditions; in this volume, Melinda Zeder, in a report distilled from an undergraduate honor's thesis, has reviewed the history of this type of study and added the results of her own investigations, but obviously we are only entering this promising field of research.

Another problem needing investigation is that of the rapidity of morphological change under varying kinds of selective factors of early domestication; one finds numerous published statements, with little accompanying data, about the numbers of generations that would be required before the skeletons of wild animals taken into domestication would be sufficiently modified to enable a modern investigator to note the differences in some of the broken bones recovered from archaeological sites. Modern controlled experiments may give us some clues as to some trends, but will never duplicate the experience of animals entering into domestication.

At present we are not even able to correlate in meaningful terms the number of bones recovered for each species with the number of animals for that species killed during the time the site was occupied, nor at present are we certain from our raw data what the relative numbers of the different species were. Until we solve this basic problem the archaeozoologist is not carying his or her share of the load as a member of the archaeological team. In an effort to gain the required interpretation, some faunal analysts depend upon a count of the number of identifiable fragments of each species, but this procedure would be valid only if each animal, no matter what its size, had been treated identically and yielded the same number of identifiable fragments; this assumption is not valid, as larger animals usually have more pieces of identifiable bones unless, as often happens, the larger animals were butchered away from the site and thus most of their bones are not recovered. Another technique is to weigh the identifiable fragments of bones of each species separately, and compensate statistically for the difference in skeletal weight; this procedure would be valid only if one can be certain that all bones of each animal were preserved and salvaged in the same proportions or that, if differences exist (as they must), we have a measure of those differences (as we don't).

The typical method of choice of faunal analysts for determining the numbers and relative proportions of animals represented by the broken bones excavated at a site is to determine the minimum number of individuals (MNI) of each species which must have been present originally to yield the bones excavated. Actually, the MNI yields a value dependent upon chance interment, chance preservation over time, and chance recovery, but these factors are often not realized. Indeed, statistical analysis has demonstrated that the MNI is valid, and then only as a measure of relative numbers of different species (never as an actual number of individuals), if the counts of identified bones are in the range of a few hundreds of thousands, far more bones than a faunal analyst usually has, or has the time to identify.

At present, some faunal analysts with statistical proclivities are attacking this problem of the relationship between the numbers of each species once being utilized by a community and the number of each kind of bone of each species which is excavated from that site at a later time. One startling conclusion from such studies is that each bone or piece thereof recovered archaeologically may well represent a different individual of the original population! Obviously, we still have much to learn, but the fact that we are learning augurs well for the future.

I have written enough, I think, to illustrate the point that faunal analysis (= zooarchaeology, osteoarchaeology, archaeozoology) is a young, vibrant, and growing subdiscipline within archaeology. My earlier conclusion re faunal analysis was that people like myself, with training in vertebrate zoology and paleontology, and fascinated with archaeology, should move into the area of overlap between zoology and archaeology, on the basis that most archaeologists of an earlier generation had neither training nor liking for biology and didn't know the difference between most animals even when viewed in the flesh. At the same time I knew then, and have since had this knowledge fortified by experience, that collections of archaeological bone, salvaged from an excavation by nonbiologists and handed over to a vertebrate morphologist who had never seen the site or often even the continent of that site, would lose most of both zoological and cultural meaning. I was right in my thinking, but cultural evolution within academia - at least in the United States - has produced an answer to the problem I had not foreseen. In general, in this country, zoologists have remained oblivious to the fascinations of zooarchaeological research, but those few who have become involved have mostly moved into Departments of Anthropology (Dexter Perkins at Columbia, Paul Parmalee at Tennessee, Stanley Olsen at Arizona, and myself at Illinois). Others with primarily zoological or paleontological training have appointments in museums, where teaching is nil or not the major requirement and freedom to pursue peripheral research is more acceptable; Barbara Lawrence at the Museum of Comparative Zoology, John Guilday at the Carnegie Museum, and Priscilla Turnbull, a free-lancer at the Field Museum, come to mind. None of these, however, are in the anthropological sections of their museums.

The interesting development in this whole field of faunal analysis in the United States, a development completely unforeseen by me, is that many graduate students specializing in archaeology in Departments of Anthropology have additionally learned the necessary zoology, often teaching themselves for the most part, to become not only capable but innovative zooarchaeologists. The spirit of adventurous bootstrap-lifing is not dead in this country, and those born with brains, ambition, and courage can still create their own frontiers.

Charles A. Reed
University of Illinois
at Chicago Circle
April 1978

PREFACE

The purpose of this symposium is to serve as a forum for discussion of different approaches to faunal analysis as dictated by differing sites, their problems, and the orientation of both excavator and analyst. Participants will be individuals presently engaged in the analysis of faunal materials from Middle Eastern sites. Emphasis will be placed on methods and approaches and not on the results of investigations. It is hoped that the inclusion of both zoologists and archaeologists trained in bone analysis will serve to promote a productive discussion of present concerns and future directions for research [Society for American Archaeology 1975, p. 16].

The papers included in this volume result from a symposium, "Approaches to Faunal Analysis in the Middle East," organized by the editors and held as part of the 40th Annual Meeting of the Society for American Archaeology (8-10 May 1975, Dallas, Texas). The expressed purpose of the symposium is stated in the quotation above. The underlying intent of the organizers was to bring together faunal analysts working in European institutions with those based in North America in order to permit these individuals of diverse backgrounds and interests to exchange information and ideas in a series of formal and informal sessions. In addition to those contributing papers to this volume, the following individuals participated in the symposium: Charles Reed (University of Illinois at Chicago Circle), Priscilla and William Turnbull (Field Museum, Chicago), and Hind Sadek Kooros (University of Tehran, Iran). The organizers particularly regret that Dexter Perkins, Jr. and Patricia Daly (Faunal Analysis Group, Columbia University) were unable to attend, the more so since, in the period intervening between then and now, we have been tragically deprived of future contributions from the latter.

Following the symposium, the editors solicited papers from the participants. Some of the contributions received faithfully reflect the original presentations while others are new or greatly modified versions which grew out of the discussions. Editing has been kept to a minimum and there has been no effort to establish a consensus. Rather we have preferred to focus on the diversity of approaches, emphases, and opinions so evident throughout the course of the meetings. This diversity has been highlighted in our introductions to each of the parts of this volume: *Methodology, Measurements, Domestication and Environment, Papers on Faunal Remains from Shahr-i Sokhta,* and *Coding Systems.*

The common denominator for the symposium and for this volume is the area of research--the Middle East--that region lying between the outer edges of the Euphrates and Indus floodplains but including also, for our purposes, the Levant and Palestine. This Palaearctic region displays marked environmental variety due to a wide range of climatic and topographic conditions and diversity of vegetational forms. It has provided, during the Holocene, habitats for more than 150 species of mammals alone including elements of both the African and Indian faunas. Among those with more specialized habitat requirements are some, including various rodent species, which occur frequently in archaeological deposits and can be used as keys to the reconstruction of past environments. Among the less specialized forms are those, including some species of artiodactyl, which were the probable ancestors of the economically most important domestic animals. This latter group is closely linked to major cultural processes in the ancient Middle East including the establishment and growth of sedentary village communities and the development and maintenance of urban life and state societies. Given these characteristics, it is natural that "the analysis of past environments" and "domestication" were two topics of particular concern at the symposium, both in the formal sessions and during informal meetings. These topics also form the underlying themes for most of the contributions to this volume although they are addressed directly only in the third section.

Another matter of some concern at the symposium also reflected in this volume was the prompt and adequate publication of faunal studies. The analysis of faunal complexes from the Middle East is still in its infancy with the material studied coming from sites spread thinly and unevenly over both time and space. Very few of these analyses have been published in detail although many have been reported in often provocative preliminary statements which were characterized by one discussant at the symposium as "very fine for the archaeologist and not too many pages to print." A partial explanation for the lack of final reports, of course, is that a number of them are "in press" and have been for years! The prompt publication of all archaeological material is highly desirable, and one can argue that should the faunal analysis be completed before the rest of the site report, it should be published immediately. One can also contend, however, that interpretation of faunal material on the basis of minimal amounts of archaeological information (chronological division only) is questionable for stratigraphically and contextually complex Middle Eastern mounds, and that should the faunal analysis be published separately, a detailed discussion of the *archaeological* limitations of the data must be included. What should *minimally* be included in a final report is a matter which was raised by Barbara Lawrence some years ago at the Budapest symposium (1971):

> The need to agree on some fundamental standard for our reports is pointed up by the propensity of synthesizers to treat different publications on fauna as of equal value and pertinence. What they cull from the literature, they use to construct theory, often in ignorance of the reliability or applicability of the statements used. Until zoo-archaeologists furnish these synthesizers with more complete and precise data, they will continue to do so. Consequently it is hoped that, at this meeting where so many of us are aware of these problems, a foundation can be laid for a uniform approach to the identification, reporting and analysis of faunal material from archaeological sites (Lawrence 1973, p. 402).

The 1975 Dallas symposium can be seen as another step in the process begun in Budapest and continued in Groningen (1974: Clason 1975) --a process of communication of individual concerns, and through such interchange, a gradual coming together of practices and procedures.* We hope that this volume, also, will serve to further those goals.

Acknowledgments

In this our first book, it gives us particular pleasure to pay tribute to the people who have helped us along the way. Our mentors C.C. Lamberg-Karlovsky and Henry Wright have been patient and supportive throughout the long periods when we were working on this and not on something else. The staff of the Publications Department of the Peabody Museum and particularly Dick Bartlett were a constant source of good advice and good cheer while we painfully pieced together the pages. We have learned early how much real work goes into putting together a book from cover to cover. Martha Smith copy-edited the original manuscripts and painstakingly proofread them again after Jean Norling, with dispatch and care, had typed final copy. We greatly appreciate the special efforts of both women on our behalf.

Fred and Martha Zeder are responsible, in large measure, for the success of the 1975 symposium and for the publication of this volume. Their gracious hospitality was central to the productivity of both the formal and informal discussions that took place in Dallas. They have also provided generous financial and moral support for the publication of these proceedings. Without their aid in all aspects of the endeavor, this volume never would have appeared.

With the enthusiastic support of all contributors, we have chosen to dedicate this collection of papers to Barbara Lawrence on the occasion of her retirement (1977) from the position of Curator of Mammals in the Museum of Comparative Zoology. Following in the footsteps of Glover Allen, she too became a pioneer of American zooarchaeology. Her great respect for the special character of zoological information and insistence on care and precision is tempered with an unusual insight into the nature of the archaeological inquiry and the strengths and weaknesses of zooarchaeological data. Her careful work, constructive criticisms, and good sense have won for her the respect of colleagues throughout the world. To one of us in particular she has been a continuing source of encouragement, good advice, and sound counsel over many years of sometimes uncertain apprenticeship. She helped us organize the symposium, carried out correspondence on our behalf during critical months when we both were in the field, encouraged us to proceed with publication, and read and reread drafts of the introductory chapters. We are truly thankful for all Miss Lawrence has contributed to zooarchaeology in general and done for us in particular.

Finally, we gratefully acknowledge the patient suffering of our respective spouses during the rather painful period when this volume was being born, and, on a lighter note, wish to thank Mamma Aysha in whose Calvert Street café the whole project was conceived.

Note
* The formation of ICAZ (International Council on Archaeozoology) at the UISPP meetings in Nice (1976) created a formal organization in an attempt to channel these efforts more effectively.

References

Clason, A.T., editor
1975 *Archaeozoological Studies*. Amsterdam/New York.

Lawrence, B.
1973 "Problems in the inter-site comparison of faunal remains," in J. Matolcsi, ed., *Domestikationsforschung und Geschichte der Haustiere*, pp. 397-402. Budapest.

Society for American Archaeology
1975 *Program and Abstracts*, fortieth annual meeting, 8,9,10 May 1975, Dallas, Texas.

Richard H. Meadow
Melinda A. Zeder
April 1978

PART ONE

METHODOLOGY

> *There is one important thing that in all of this must be stressed to all archaeologists, not to members of this group, but to others, and that is that the zooarchaeologist must be an active member of the group, cooperating in the field. There still is the tendency too often to have the bones or what remains of them picked out by the excavator and brought back to some zoologist and dumped on his desk...And I think that is the very worst way to accomplish anything [Charles A. Reed, 9 May 1975].*

The three papers included in this section are by authors who have worked actively with archaeologists excavating in the Middle East. Their contributions show both a commitment to such participation and a realistic understanding of the nature of the archaeological undertaking and of the particular characteristics of material recovered from excavations. LaBianca's team approach to faunal analysis in the field is unique in the archaeology of this area and reflects the author's special talent for organizing untrained personnel. Large quantities of data can be recorded quickly and, given proper supervision, accurately with a resulting potential for large-scale contextual and distributional analysis of the faunal remains. Such an approach, however, is only as strong as the archaeology on which it must be based; the methods often used to clear large areas of complexly stratified Middle Eastern *tells* may render much material unsuitable for contextual analysis. Therefore, any person undertaking such studies must have a detailed understanding of the archaeology or must be able to work closely with those who do.

Another feature characteristic of most projects in the Middle East is a limited supply of both time and money. As Lawrence points out, this reality plays an important part in determining the extent to which detailed studies like scrap analysis can be carried out. Each analyst must decide on the basis of experience and interest what trade-offs can be made given the nature of the material and the resources at hand. In this instance an understanding of the techniques of excavation and conditions of recovery can serve the analyst well by helping him or her to determine the appropriateness of particular kinds of study. No matter what choices are made, however, the procedures used in recovery, sampling, and analysis must be made explicit in the publication of faunal data.

Meadow's paper raises another issue. Faunal remains are usually reported in units based on chronological divisions. The various material so described are then compared with each other with the intent of determining changes in faunal exploitation patterns over time. The underlying assumption is that the remains from one period present a good "average" of the cultural activities of that period. All archaeological remains, however, have passed through various cultural and natural filters before being recovered. The fact that differences in these filtering processes may affect the configuration of the eventually recovered materials is only now becoming appreciated. The nature and consequences of processes of disposal, deposition, and preservation are topics of some interest to paleontologists as well as to archaeologists with the result that taphonomic studies and ethnoarchaeology are becoming increasingly important to their respective disciplines (e.g., Behrensmeyer 1975; Yellen 1977). Another source of bias is, of course, incomplete recovery; this aspect is explored in the preliminary remarks to the Shahr-i Sokhta papers in Part Four.

References

Behrensmeyer, A.K.
  1975   *The Taphonomy and Paleoecology of Plio-Pleistocene Vertebrate Assemblages East of Lake Rudolf, Kenya.* Bulletin of the Museum of Comparative Zoology, vol. 146, no. 1.

Yellen, J.E.
  1977   *Archaeological Approaches to the Present.* New York.

R.H.M.
M.A.Z.

THE LOGISTIC AND STRATEGIC ASPECTS OF FAUNAL ANALYSIS IN PALESTINE

Øystein Sakala LaBianca, Andrews University and
American Schools of Oriental Research
Expedition to Tell Ḥesbân, Jordan

Introduction

After four seasons in the field as a faunal analyst, I have come to regard the subject of this paper--the logistic and strategic aspects of faunal analysis in Palestine--as critical for the success of faunal analysis at the particular site with which I have been associated, Tell Ḥesbân in Jordan. First, a few words about the dig itself.[1]

Since 1968 five seasons of archaeological excavations have been carried out at Tell Ḥesbân, biblical Heshbon, located fifteen miles southwest of Amman, the capital city of Jordan (Boraas and Horn 1969, p. 97). The site, which is well known from literary sources of antiquity (Vhymeister 1968), was selected for archaeological investigation because it looked as if it could produce data relevant to questions about the early history of Jordan and the possible conquest and later settlement of Israelites in this region (Boraas and Horn 1969, pp. 99-103).

Excavations, following the principles of the Wheeler-Kenyon method, of seven areas on and about the acropolis and in an ancient cemetery have unearthed remains from the Islamic, Byzantine, Roman, Hellenistic, Persian, and Iron Age periods. The site, therefore, has an occupational history beginning about the twelfth century B.C. and ending around the fourteenth century A.D. The modern Arab village of Ḥesbân, which has a population of about 800 and is located adjacent to the archaeological site, has been in existence only since the turn of the century.

The excavations at Tell Ḥesbân, which were begun by Siegfried H. Horn and are currently directed by Lawrence T. Geraty, both of Andrews University Theological Seminary (Berrien Springs, Michigan), have typically involved a large number of participants. For example, the 1976 campaign involved 240 persons in all: 100 foreigners and 140 Arab workmen. About 1200 animal bone fragments were unearthed daily in 1976 by the 20 simultaneous stratigraphic operations at the tell. Preliminary reports have been published following each season summarizing the major accomplishments of each stratigraphic operation (Boraas and Horn 1969, 1973, 1975; Boraas and Geraty 1976, 1978). These reports include discussions of the organization of the faunal analytical work as well as findings based on completed studies of selected portions of the animal bone corpus (LaBianca 1973a; LaBianca and LaBianca 1975, 1976; Boessneck and von den Driesch 1978, see also 1977; Boessneck 1977; Crawford 1976). The pertinence of faunal analysis to illuminating the history of animal exploitation at Tell Ḥesbân, and its pertinence to the more general problem of discovering the underlying psychological, cultural, and ecological principles which explain the course of that history from ancient times down to the present, has been discussed elsewhere by the writer (1978, in press). The same writings account for the writer's initiation of ethnographic, taphonomic, and meteorological studies in the present-day village of Ḥesbân (1976, 1978) and for his promotion of other kinds of environmental studies (Crawford and LaBianca 1976; Crawford, LaBianca, and Stewart 1976; Alomia 1978).

Skeletal Part Assemblages versus Findspot Assemblages

It is when insufficient funds or government regulations prohibit the shipment of large quantities of animal bones to distant lands for future study that the challenge for the faunal analyst at a large dig comes into focus. This situation establishes limits to the time available for processing the bone data. It also establishes a potential crisis relative to laboratory work space and equipment necessary for faunal analysis. Finally, this situation generates a critical need for the establishment of priorities about how much and what kinds of data to collect. This question, in turn, is inextricably tied to questions of strategy.

Consider, for example, the important issues which are involved in deciding whether or not bones are best analyzed in association according to archaeological findspot or in assemblages grouped according to element. In the following discussion the former will be referred to as "findspot assemblages" and the latter as "skeletal part assemblages."

The strategy followed at Tell Ḥesbān has involved analysis using skeletal part assemblages. The reason for using this strategy is best explained with reference to the archaeological situation at Tell Ḥesbān and to the arrangements for analyzing the bone data. At this site the faunal analyst cannot assume that all findspots designated as Early Roman by the archaeologists are equally Early Roman. There are Early Roman fills from pits, cisterns, or foundation trenches which were accumulated during the Early Roman Period, but which contain artifacts and bones from one or more earlier periods as well; there are sealed floors, cisterns, and tombs with exclusively or predominantly Early Roman artifacts or bones; there are Early Roman installations such as ovens, wine presses, and baths with unusual accumulations of artifacts and bones, not necessarily from the same period. Each campaign brings to light dozens of each of these various kinds of deposits, complicating enormously the popular layer-cake notion of the structure and composition of a tell.

This situation requires that the faunal analyst adjust his strategy to cope with both the particulars of the archaeological context and with the constraints upon shipment home of the samples (mentioned above). Such a requirement creates a dilemma, however, in that typically the particulars of any archaeological context are only <u>partially</u> interpreted by the archaeologists in the field, and thus there is usually insufficient data available at the time of the analysis to determine which findspot assemblages should be focused upon. Consequently, each and every skeletal part must be described with respect to <u>all</u> of the attributes which are regarded as informative for such questions as the minimum number of individuals of a species, the age of the animals at time of death, and so on, so that at a later date, when all the pertinent particulars of the archaeological context have become known, the bones can be reassembled, using a computer, into analytically suitable assemblages.

These conditions then, are one reason why, at Tell Ḥesbān, our strategy has involved analysis of skeletal part assemblages rather than findspot assemblages, although the latter strategy was not bypassed entirely, as will be seen later. Another reason is that skeletal part assemblages lend themselves more readily to teamwork between specialists and nonspecialists, so that all the important decisions regarding the taxonomic, anatomical, physical, and cultural attributes of individual bone fragments are made by the specialist, while the enormous amount of preparation including cleaning and labeling and follow-up work including recording and data entry can be carried out by nonspecialists. Before discussing the logistic arrangements needed to carry out this strategy, however, a summary of some of the issues involved in deciding whether to employ a findspot assemblage strategy or a skeletal part assemblage strategy seems appropriate.

A) Whereas analysis involving skeletal part assemblages requires that all bones be individually labeled, this is not essential when dealing with findspot assemblages insofar as the bones are kept together.

B) Whether splinter fragments belonging to the same bone are more likely to be discovered and restored to each other in one or the other strategy is debatable. On the one hand, intuitively it would seem that fragments belonging to the same bone are most likely discovered in findspot assemblages; that is, if one assumes either that the splinters belonging to the same bone were buried within the space designated by the excavator as the findspot, or that the splintering of bones happens, for the most part, after they have been buried, as the result of postdepositional processes in the ground or through rough handling by the excavators. In either case, chances are that findspot assemblages offer the best opportunity for restoration.

On the other hand, experience with skeletal part assemblages has shown that once all splinter fragments belonging to the same category--for example, all tibia of sheep--have been assembled, splinters belonging to the same bone are readily discovered and restored, with the advantage that splinters from different findspots which belong together are discovered and can be restored. Clearly, this is an issue which needs to be investigated systematically in the future.

C) Bones belonging to the same skeleton are usually kept together in findspot assemblages, whereas they may inadvertently be permanently separated in skeletal part assemblages, although this depends, of course, upon the kind of effort that is made to separate out bones which belong together before the bones are sorted into skeletal part assemblages.

D) Detailed descriptions of individual fragments can be recorded on standardized forms and/or entered into the computer with greater accuracy and efficiency when dealing with skeletal part assemblages because there are usually far fewer different attribute combinations involved. For example, all right distal fused ends of sheep humeri are recorded and/or entered together instead of having to be entered as they occur sporadically in findspot assemblages.

E) The unskilled assistance of hired workers and volunteers and the semiskilled assistance of trainees are more readily utilized in dealing with skeletal part assemblages because, as was pointed out in "D" above, far fewer technical decisions need to be made by the assisting personnel.

Four Phases of Faunal Analysis at Tell Ḥesbān

The process whereby animal bones excavated from the earth are transformed into evidence relative to ancient faunal exploitation practices can be thought of as entailing four distinct phases.

The bone excavating phase. This phase occurs in all stratigraphic operations on the tell and embraces all activities having to do with the way bones are excavated, collected, and transported from excavation site to the faunal analysis station on the tell.

The pre-analytical phase. This phase occurs at a special field station on the tell--the bone tent. It entails initial sorting of the bones, cleaning, labeling, weighing, and counting of the bones in the field.

The analytical phase. This phase occurs in a

laboratory at headquarters and embraces all activities having to do with identifying, categorizing, and recording the bone data.

The interpretive phase. This phase takes place following the actual fieldwork during the ensuing months and it includes such activities as electronic data processing, data reduction, interpreting the data, and preparing the material for publication.

Whereas the logistic aspects of this four-phase process entail considerations relative to the kinds of physical facilities, equipment, and supplies involved, the strategic aspects have to do with the kinds of tasks that have to be accomplished, the sequence in which they are done, and who accomplishes them. Logistic arrangements will be discussed primarily with reference to what is required in the field, while the strategic arrangements will be discussed relative to the pertinent arrangements both in the field and at home.

Logistic Arrangements

I shall begin by detailing the logistic arrangements, first in the bone tent, and then in the bone laboratory.

The bone tent, where the pre-analytical tasks are done, contains: A) bone-drying stalls built of large rocks providing a place for drying the fragile, newly excavated bones overnight and protecting them from exposure to inadvertent human shuffling; B) sieves, with one-quarter inch mesh, which are used for holding the bones when selecting those to be cleaned and not cleaned, and when cleaning them; C) a small worktable used when weighing and labeling bones; D) a metric scale for weighing cleaned and not-cleaned bones from each findspot.

The bone laboratory at headquarters, where the analytical tasks are done, contains: A) large and medium-sized tables--never fewer than two, usually six to eight; B) an osteological comparative collection consisting mostly of bones of the domestic species found in Jordan today and assembled as a result of surveying modern settlements and picking up weather-prepared whole bones or partial skeletons (see LaBianca and LaBianca 1975); C) a small library of books and photocopied articles dealing with faunal analysis (see Toplyn 1975, p. 28); D) osteometric equipment including an osteometric board for measuring the lengths of large bones, large external and small general purpose calipers, and a metric scale for weighing bones; E) about 100 "Kellogg trays" used as sorting trays in the laboratory (these trays were constructed from discarded cereal boxes); F) miscellaneous supplies and equipment such as plastic sandwich bags for packaging unusual bones (the more, the better!), paper bags for circulating around the tell to pick up bones in, large plastic bags for storage of bones following analysis, an assortment of dental picks, rulers, scrub-, paint-, and toothbrushes, scissors, red ink pens, pencils, Duco Cement tubes, fine-point felt pens and various coding forms.

Strategic Arrangements

As was pointed out earlier, strategies involving skeletal part assemblages lend themselves to the employment of unskilled and semiskilled personnel. The specific ways in which such personnel can be employed to the advantage of the faunal analytical process at a large dig such as Tell Ḥesbân will be illustrated in the following discussion of the strategic arrangements which were followed during the 1976 campaign. I shall first describe the tasks pertaining to each of the four phases and thereafter discuss how the tasks were distributed among the personnel.

The tasks of the bone excavation phase include:
Excavating the bones. This task is accomplished as part of the stratigraphic operation on the tell. Bones are excavated along with pottery and other artifacts, but they are placed in specifically designated paper bags which are labeled with the appropriate findspot information. (See LaBianca 1978 for a discussion of the excavation of a test square for the purpose of ascertaining the adequacy of existing stratigraphic methods for the collection and preservation of faunal remains.)

Gathering in the bone bags. To ensure proper labeling and safe transportation of bone bags from the excavation site to the bone tent, bone bags are gathered and taken to the bone tent several times each day by the individuals assigned to operations in the bone tent.

Excavating unusual bone deposits. When unusual bone deposits were discovered by the various square supervisors, the faunal analyst or one of his full-time assistants would be called to assist with excavating the deposit. This arrangement ensured that the bones would be handled properly and that in-situ assessments could be made by the faunal analyst.

The tasks of the pre-analytical phase included:
Selecting and counting the bones to be cleaned and not cleaned. The decision as to which bones to clean and which bones not to clean was made in accordance with the same considerations used in previous seasons for deciding which bones to save and which ones to discard. Thus fragments which could be identified as skull fragments, ribs, vertebrae, long bones showing articulating surfaces, and whole bones were cleaned. The rest were placed in small paper bags, labeled, and returned to the bag from which they originally came, along with the bones to be cleaned.

Cleaning the bones. This task entailed "dry-brushing" the bones to be cleaned using scrub- or toothbrushes and dental picks. Some restoration of fragments was also done as the bones were cleaned.

Labeling the bones. Bones were labeled, using a fine-point, black ink felt pen, according to their stratigraphic place of origin. When the bones were too small to be labeled individually, they were placed in a plastic sandwich bag and labeled collectively.

Weighing and counting the bones. The contents of both bags, the one with the cleaned bones in it as well as the one inside it with the not-cleaned bones, were counted and weighed separately and the resulting information was recorded on a specifically designed computer-oriented data sheet.

Consolidating the contents of bags from the same findspot. Since bones from the same findspot were usually collected over several days, more than one bag was usually turned in by the time a particular findspot had been fully excavated. Consequently it was necessary to consolidate the contents of the bags coming from the same findspot. This was done back at headquarters.

The tasks of the analytical phase included:
Bone reading. Bones were laid out on tables in the bone laboratory according to findspot. They were then "read," i.e., the faunal analyst identified and counted the bones of each of the domestic species represented in each findspot assemblage, and this preliminary information was recorded by the respective square supervisors in their field notebooks.

Obtaining contextual information for each findspot assemblage. A specifically designed computer-oriented "Contextual Data Information Form" was filled out for each findspot assemblage in conjunction with the bone reading. The respective square supervisors were asked to provide the faunal analyst with available information about the dating of the assemblage, based on associated pottery and artifacts; the stratigraphic particulars of the deposit involved; and interpretations as to the original purpose(s) of the deposit, i.e., whether it had been a floor, foundation trench, cistern, wine press, and so on.

Findspot assemblages sorted according to major historical periods. On the basis of the information obtained using the Contextual Data Information Form, the findspot assemblages, which until now were in no particular order, were grouped together according to major historical periods. This grouping, however, was at best preliminary and did not take into account the many particulars of the archaeological context which have a bearing on how these assemblages eventually are combined and analyzed, as discussed earlier.

Findspot assemblages sorted according to species. Cleaned and not-cleaned bones from the same findspot assemblage were sorted according to species. The bones of each species were weighed and the weights were recorded on a specifically designed computer-oriented data sheet.

Findspot assemblages sorted according to skeletal parts. Both cleaned and not-cleaned bones were sorted according to skeletal parts. However, while the cleaned bones were distributed according to skeletal parts and species into Kellogg trays, the not-cleaned bones were first reduced to only splinters of sheep/goat, then promptly described on a specifically designed form (provided by Drs. Boessneck and von den Driesch) for recording the occurrence of the various skeletal parts in findspot assemblages. Thereafter the not-cleaned bones which were not sheep/goat splinters were cleaned and labeled while the rest of the not-cleaned bones were returned to their bags and packed away. Unusual bones were also separated out at this point and stored separately in plastic bags, while measurable bones were individually numbered so they could be culled out for measuring at a later point (described below).

Skeletal part assemblages sorted according to anatomical criteria. Skeletal part assemblages comprising the bones of a particular species were next sorted according to anatomical criteria: right or left; proximal or distal; fused, unfused, or in fusion; and so on for the long bones and, as seemed appropriate, for skull fragments, vertebrae, ribs, and the like. (See LaBianca and LaBianca 1975 for a more detailed description of the categorization scheme employed at Tell Ḥesbân in 1973.)

Sorted skeletal part assemblages described according to physical and cultural characteristics and coded. Sorted skeletal part assemblages were individually described on "tickets" specifying the appropriate computer-oriented codes and where these codes should be coded on the specifically designed computer-oriented coding sheets. Individual assemblages were then queued up for coding. Once they reached the coding personnel, the individual bones in each skeletal part assemblage were scrutinized for signs of physical erosion, rodent gnawing (usually circled by the faunal analysts), and the like, measured for maximum length on a "bone-scale" (consisting of an 8-1/2 by 11 inch sheet of paper with lines drawn across at 5 mm intervals and with incrementing numbers--1-54--at the end of each line), and, together with the information contained on the bone "ticket," the appropriate codes were entered on the data sheet. When a sorted skeletal-type assemblage had been thus processed, it was again queued for verification by a coding supervisor. Once verified, the bones which were slated for measuring were separated out, and the rest were bagged in plastic bags and packed away.

Measuring. Bones slated for measuring were returned in element groups to the faunal analysts who then measured them following the scheme for measuring animal bones described by von den Driesch (1976). The measured bones were then likewise bagged and packed away.

Although the tasks of the interpretive phase are still in mid-stream, certain strategic arrangements pertaining to this phase are well established, particularly those pertaining to how the computer will be utilized.[2]

Data-entry and verification. Until recently, these tasks have involved key-punching of the data on 80-column punch cards and subsequent verification of the data by the computer using a specifically designed verification program. Currently an arrangement is being tested whereby the data is entered on a teletype terminal which is on-line with the computer which, in turn, has been furnished with an interactive data-entry and validation program so that each record is validated as it is being entered and automatically stored in a designated file.

Data file linking. Since the information pertinent to interpreting the bone data--and even the bone data itself--are being entered into the computer as separate files, a specifically designed linking program is being used for linking certain of the data files. For example, the file which contains the information about the anatomical, physical, and cultural attributes of each individual cleaned bone fragment is being linked to the file which contains the information about the occurrence of not-cleaned splinters of sheep/goat, and the resulting file is, in turn, linked to the contextual information file so that each bone in the 1976 season's bone corpus of 42,000 fragments is individually linked to contextual information about the deposit from which it came.

Isolating analytically appropriate assemblages. Existing computer programs allow us to retrieve from the set of all animal bone records any subset we may require, and to create permanent subfiles. Thus, our strategy involves isolating subfiles consisting of the bones from each individual historical period. Each of these subfiles, in turn, will be further dissected according to the particulars of the various archaeological contexts within the individual historical periods. In this manner, it is possible to ensure that the bones which will be subjected to further statistical analysis come from well-defined deposits, and hopefully, as a result, the conclusions which are drawn will have more integrity.

Statistical analysis of isolated assemblages. Isolated assemblages will be statistically analyzed, with some assistance from the computer, to ascertain the relative importance of the various species; particulars about the natural history and cultural use of each individual species; and particulars about the depositional and postdepositional events which have a bearing on the interpretation of each isolated assemblage.

Whereas the tasks of the excavation and preanalytical phases as well as the first three tasks of the analytical phase were carried out each day during the seven weeks of the dig, the last five tasks of the analytical phase were carried out during a four-week post-season bone lab (LaBianca 1978). During the 1976 season, two full-time assistants with on-the-job training were responsible for all the tasks of the bone excavation and pre-analytical phases. Each had occasional volunteers to supervise--one on the tell in the bone tent and the other at headquarters in the bone lab. The full-time assistant in the bone lab was also responsible for laying out the bones according to findspot assemblages in preparation for bone reading and for obtaining from the archaeologists the contextual information called for by the Contextual Data Information Form. Bone readings were done by one of the faunal analysts each afternoon of the work week. After the bone reading, the bone lab assistant would sort the findspot assemblages according to major historical periods.

During the first two weeks of the post-season bone lab, Drs. Boessneck and von den Driesch sorted according to species and skeletal parts the 42,000 bones from 1976 as outlined in the analytical phase. During the third and fourth week they measured all the bones from the 1976 season, in addition to which they went through approximately 20,000 bones from earlier seasons, identifying unusual fragments and measuring all measurable bones.

Categorization of the skeletal part assemblages according to fragment types required all four weeks--the writer working the whole period, while supervising the bone lab, and Mike Toplyn working two weeks. Their categorizations were routinely verified by either Dr. Boessneck or Dr. von den Driesch. All bone "tickets" were prepared by the writer.

Four full-time assistants with on-the-job training worked during the entire four-week period in order to check the 19,602 cleaned and ticketed bones for physical and cultural characteristics and describe them individually on coding sheets. It should be noted, however, that these assistants worked, on the average, eight hours five days each week, while the specialists mentioned above worked, on the average, twelve hours each day.

During the 1976 season, full-time assistants and volunteers contributed an estimated 1400 person hours in carrying out their tasks during the first three phases, while the four specialists contributed an estimated 700 person hours to the analysis of the bones from the 1976 season. Thus 42,000 bone fragments were processed in altogether 2100 person hours--fully 2/3 of which is accounted for by assisting personnel. The amount of time spent on processing one bone fragment through the first three phases was about 3 minutes, the same time as has been reported for the 1974 season (LaBianca 1975, p. 6).

Reflections on Strategy

Two fundamental issues in faunal analysis come into focus as I reflect upon the strategic arrangements described here. The first has to do with the issue of how to regard individual bone fragments for the purposes of analysis and interpretation, the second has to do with the issue of standardization of information in faunal analysis.

As was pointed out in the beginning of this paper, the strategy at Tell Ḥesbân has been to make as few assumptions as possible in the field about which findspot assemblages are the most suitable for analysis, given the complex archaeological situation at this site. As a consequence of this, each individual bone was regarded as if it were as valuable as any other bone for the purposes of statistical interpretation. This, in turn, magnified considerably the amount of categorization and recording which had to be done, despite the advantages in terms of efficiency and accuracy resulting from following a skeletal part assemblage strategy.

Much less categorization, recording, and handling of the bones by nonspecialists would have been necessary could we have safely followed a findspot assemblage strategy. Such a strategy was not used at Ḥesbân, as was noted previously, because it would have required reliance on only partially synthesized findspot information and would have had to have been carried out almost entirely by specialists, thus requiring more time than was available in the field. A findspot strategy when used with an attribute recording system requires that each bone be described with respect to all pertinent attributes as it is come upon in the course of analyzing the assemblage from one findspot, a task which cannot be carried out by nonspecialists.

It is conceivable that, in the future, criteria will be delimited for determining whether or not a given archaeological deposit contributes statistically meaningful bone assemblages. Such criteria could then be used at future digs for selecting which findspot assemblages to concentrate on, and thereby reducing considerably the work needed in order to attain reliable generalizations based on faunal analysis.

Much effort has gone into standardizing procedures and categories for use in describing bone fragments at Tell Ḥesbân. This, in fact, has been a critical prerequisite for implementing the strategy described in this paper, because, for tasks to be delegated to unskilled assistants and for the millions of bits of information to be processed using the computer, unambiguously delimited categories and classifications, which could be explained to others, had to be drawn up. Not surprisingly, therefore, those tasks which involved the use of unambiguous classifications were accomplished with better precision by assistants than those which involved ambiguous classifications, as was the case, for example, with the classification of cultural and physical marks, an area where much improvement is still needed.

One danger with standardization, however, is the constraints it establishes with regard to ways of doing and seeing things. This stagnation effect of standardization can be checked, however. Thus, at Tell Ḥesbân, procedures have changed somewhat from season to season (compare LaBianca 1975), and the data categorization and coding system have been improved each season. In fact, the computer programs currently in use have been specifically designed so that changes in the kinds of data being entered and in how it is analyzed can be made whenever it is deemed necessary and without much trouble.

Finally, significant advances in faunal analysis of bones from large expeditions in the Middle East require higher levels of data integration and control. This, in turn, signals the need for improved coordination of the activities of the participating staff, especially with regard to establishing the archaeological particulars of bone deposits.

Notes

1. An abbreviated version of the paper which I read at the Dallas Symposium was published in the Newsletter of the American Schools of Oriental Research (LaBianca 1975). This original version dealt with my experience in the field through the 1974 campaign at Tell Ḥesbân. The present paper incorporates the insights and experiences gained during the 1976 campaign, when I had the opportunity of working with Drs. Joachim Boessneck and Angela von den Driesch of the Institut für Palaeoanatomie, Universität München. These two specialists concentrated their efforts, during the month of August, upon the identification and osteometric analysis of faunal remains from the 1976 season and earlier seasons, enabling the writer to concentrate on the overall logistic concerns described in this paper. Many of the ideas and insights which are new with this paper are attributable to the good cooperation and constructive criticisms offered me by Drs. Boessneck and von den Driesch.

My attendance at the Dallas Symposium was made possible financially by Dr. Lawrence T. Geraty, Director, Andrews University Heshbon Expedition. Others to whom I am indebted for research grants and other assistance toward support of the faunal analytical work during the 1976 campaign are the American Schools of Oriental Research, the Deutschen Forschungsgemeinschaft, the Earthwatch Research Associates, and the Heshbon Expedition Fund.

2. The computer programs which are currently used for processing the bone data are the result of collaboration between Paul Perkins, of the Institute for Informatics Research and Computer Design, Lexington, Massachusetts, who wrote the programs and provided computer facilities, and the writer, who systematized the procedures involved in faunal analysis so they could be programmed for the computer, and who helped test the programs.

References

Alomia, K.
   1978   "Notes on avifauna of Ḥesbân," *Andrews University Seminary Studies,* vol. 16.

Boessneck, J.
   1977   "Funde vom Mauswiesel, *Mustela nivalis* Linné, 1766, auf dem Tell Hesbon, Jordanien," *Säugetierkindliche Mitteilungen,* vol. 25, no. 1, pp. 44-48.

Boessneck, J. and A. von den Driesch
   1977   "Hirschnachweise aus frühgeschichtlicher Zeit von Hesbon, Jordanien," *Säugetierkundliche Mitteilungen,* vol. 25, no. 1, pp. 48-57.
   1978   "Preliminary analysis of the animal bones from Tell Ḥesbân," *Andrews University Seminary Studies,* vol. 16.

Boraas, R.S. and L.T. Geraty
   1976   "The fourth campaign at Tell Ḥesbân," *Andrews University Seminary Studies,* vol. 14, pp. 1-216.
   1978   "The fifth campaign at Tell Ḥesbân," *Andrews University Seminary Studies,* vol. 16.

Boraas, R.S. and S.H. Horn
   1969   "The first campaign at Tell Ḥesbân," *Andrews University Seminary Studies,* vol. 7, pp. 97-216.
   1973   "The second campaign at Tell Ḥesbân," *Andrews University Seminary Studies,* vol. 11, pp. 1-144.
   1975   "The third campaign at Tell Ḥesbân," *Andrews University Seminary Studies,* vol. 13, pp. 101-248.

Crawford, P.
   1976   "The Mollusca of Tell Ḥesbân," *Andrews University Seminary Studies,* vol. 14, pp. 171-176.

Crawford, P. and Ø.S. LaBianca
   1976   "The Flora of Ḥesbân," *Andrews University Seminary Studies,* vol. 14, pp. 177-184.

Crawford, P., Ø.S. LaBianca, and R.B. Stewart
   1976   "The Flotation Remains," *Andrews University Seminary Studies,* vol. 14, pp. 185-188.

Driesch, A. von den
   1976   *A Guide to the Measurement of Animal Bones from Archaeological Sites.* Peabody Museum Bulletin 1. Cambridge (Harvard University).

LaBianca, Ø.S.
   1973a   "The zooarchaeological remains from Tell Ḥesbân," *Andrews University Seminary Studies,* vol. 11, pp. 133-144.

1973b	"A study of post-cranial remains of sheep and goat from Tell Ḥesbân, Jordan," unpublished manuscript, Harvard University, Anthropology 207 (May 24, 1973).	
1975	"Pertinence and procedures for knowing bones," *Newsletter of the American Schools of Oriental Research,* no. 1.	
1976	"The village of Ḥesbân: an ethnographic preliminary report," *Andrews University Seminary Studies,* vol. 14, pp. 189-200.	
1978	"Man, animals, and habitat at Ḥesbân, Jordan: an integrated overview," *Andrews University Seminary Studies,* vol. 16.	
in press	"The diachronic study of animal exploitation at Ḥesbân," in L.T. Geraty, editor, *The Archaeology of Jordan and Other Studies.* Berrien Springs: Andrews University Press.	

LaBianca, Ø.S. and A.S. LaBianca
- 1975 "The anthropological work," *Andrews University Seminary Studies,* vol. 13, pp. 235-247.
- 1976 "The domestic animals of the Early Roman Period at Tell Ḥesbân," *Andrews University Seminary Studies,* vol. 14, pp. 205-216.

Toplyn, M.
- 1975 "Some aspects of the zooarchaeological operation at Tell Ḥesbân, Jordan," unpublished manuscript, Andrews University Heshbon Expedition, 1974 campaign.

Vhymeister, W.
- 1968 "Heshbon's history from literary sources," *Andrews University Seminary Studies,* vol. 6, pp. 158-177.

ANALYSIS OF UNIDENTIFIABLE BONE FROM ÇAYÖNÜ: AN EARLY VILLAGE FARMING COMMUNITY

Barbara Lawrence, Museum of Comparative Zoology
Harvard University

As faunal studies have become more sophisticated in the last twenty-five years, pressure has increased on archaeologists to save every fragment of bone, no matter how small and whether identifiable or not. While the insistence on total recovery has increased, the rationale for such time-consuming efforts has not. Just what we plan to tell the archaeologist in return for his effort is not necessarily clear, nor have we made clear whether the information derived from unidentifiable bone justifies the delays in excavation which are an inevitable consequence of total recovery.

An effort to help clarify these problems was made at Çayönü during the seasons of 1964, 1968, and 1972.[1] Different methods of treating unidentifiable bone, here called scrap, were tested for their possible value in producing useful information, and this paper is not so much a statement of results as a consideration of what can be found out from scrap and how.

One of our aims at Çayönü was to test the feasibility of identifying material and recording pertinent data at the site rather than shipping the entire collection to a laboratory for future study. With this in mind, the initial processing of the bone was similar in all three years, though subsequent treatment of the scrap differed. Partly this second stage depended on what had been learned in a previous season and partly on whether or not an assistant was available to help with the bone.

My efforts during the first season were limited to qualifying and quantifying the scrap on a lot by lot basis. The purpose was to see if there was any pattern in the occurrence of fragments of precise sizes or shapes which might suggest any use of the fauna other than as food. We also hoped that we might find some correlation between unidentifiable and identifiable fragments of different sized animals and thus get additional evidence for their relative importance in the diet.

As far as the first was concerned, we did indeed find enough of an apparent patterning of types of scrap to suggest for the following season a more detailed analysis of all fragments to see what proportion had been modified by use or retouch and to what extent and in what way types of fragments could be categorized.

As far as the second was concerned, the task of accurately distinguishing between fragments of bone from large, medium, or small animals was too difficult. While extremes of size, as between aurochs and small artiodactyl, are easy enough to recognize, there is continuous overlap between such intermediate forms as Cervus, Equus, Sus, Dama, and large male Capra. Since we had no comparative material to determine range of variation in thickness of bone from different parts of the shaft, not only for different species but also for different elements of the same species, identifications would have been highly subjective. Attempts to eliminate bias, by allocation to one group or another on the basis of the percentage of occurrence of identifiable bone, would have been useless as a basis for investigating differences in utilization of different species.

Nevertheless, it was clear that some useful information can be derived from such a size sort. An abundance of fragments too large to be anything but big cattle helped to confirm the relative importance of Bos. Further, if such fragments, together with identifiable pieces of limb bones, are present, it would suggest a dismembering of the carcass near the site. The value of such deductions depends of course on the certainty with which one can distinguish between Bos and Cervus.

Additionally, comparisons of Çayönü scrap with that from Jaguar Cave in North America showed that in some instances size and abundance of scrap differs sharply from site to site. To understand the reason for this, a method of sorting and recording scrap, which would make meaningful intersite comparisons possible, needed to be developed.

As a result of our work during 1964, a more precise analysis was undertaken during the 1968 season by Carolyn L. Wright. After relative amounts of identifiable bone and of scrap were measured, the scrap was then further sorted into a variety of categories and the quantity in each determined. The categories were based on size of fragment, not of animal from which they were derived, and consisted of: splinters up to 3 cm, splinters larger than 3 cm, shaft fragments, split shafts, vertebrae, ribs, scapulae, tooth fragments, possibly utilized, and miscellaneous. Our purpose was to look for and study variation in amount and kind of scrap by level/area, and to determine whether or not scrap was being utilized at the site and in what ways. Significant results were achieved for both purposes.

Referring to the first, a number of interesting points were demonstrated. Differences occur in the amount of bone recovered from different areas, and such differences have more meaning if one establishes what type and size of scrap are associated with what areas. For instance, not only is scrap more abundant in the upper levels where we have evidence of domestication but also the percentage of small to large fragments is about three times as great as in the deeper levels. Similarly, comparisons of scrap with identifiable bone and with architectural features yield such information as a relative scarcity of scrap compared with worked bone within a number of buildings, and, within another complex a scarcity of both scrap and identifiable bone. When scrap abundance is regularly recorded, comparisons of this sort with artifacts recovered can also be made.

While much information can be recovered in the laboratory from bags of bones, the significance of scrap abundance is best studied if the zooarchaeologist actually participates at the site. In 1968, it was possible to have someone at the excavation daily to note actual scatter of the bones and the following observations will show what this method accomplished. Centrally, on the mound in the clearance called SBI, where bone was relatively abundant, it was found to be rather thinly scattered vertically as well as horizontally in an area west of the main architectural features. This distribution was also haphazard as far as kind of animal and type of bone was concerned. There was no association with floors, and the predominance of small fragments was too great to be the residue of butchering and food use. In one small part of the area, we found that partial rotting and breaking in situ of bones in rocky soils, which alternately are very wet and hard-baked dry, had resulted in individual bones being shattered and the cracks, between the often very small fragments, stained and weathered. When undisturbed, this mosaic appeared as a single, rather large, piece of bone. The fragments are easily scattered, however, and when disassociated are similar to the scrap found commonly in the upper levels of the rest of the site. All of this suggests that, in SBI, the fragments recovered during the excavation are the result of breaking of bone as it lies buried and a subsequent dispersal of the resultant pieces. If this is true, then the degree to which scrap in a given area is from different and unrelated bones seems to indicate that the bone is not primary residue of occupation but evidence of refill. Further, while patterns of bone breakage have still not been sufficiently studied, Hind Sadek-Kooros's work (1972, 1975) and a preliminary comparison with her scrap from Jaguar Cave show that this Çayönü scrap is quite unlike that caused by purposeful breaking or found in an accumulation of the residue of such breaking. Comparisons such as these substantiate the thesis that scrap analysis can produce different types of useful information.

A more thorough analysis was undertaken of the material which might have been modified by use. In the original scrap sort, about 1200 pieces were set aside as possibly modified. Of these, a more careful scrutiny under 5 to 10 power magnification showed that about three-quarters had evidence of secondary modification. The criteria used were rubbing (which differed from accidental) and chipping (as distinct from shattering caused by breaking). None of the modified fragments were finished tools, and the modifications though real are slight. Handy fragments were apparently put to use and the variety of types of wear suggests a variety of uses. Carolyn L. Wright, as an archaeologist, made a tentative typology of the different pieces. More points, 50.5%, were used than any other form, and parallel-sided pieces, rubbed on one edge, accounted for about 23%. It is also of interest that the small splinters were less frequently used than the large ones of 3 cm or over. These results have not been published, since this was a pilot project and needs further testing. Studies of similar modifications of identifiable bone should be made, and a redefining of the categories used, based on their applicability to scrap from other seasons.

During the last season, 1972, with more bone and less help, scrap analysis was limited to a consideration of the quantitative relation between scrap, seed, and soil removed in an area dug to sterile soil. A comparison of MNI (minimum number of individuals) with numbers of different types of seed found was generally confirmed by a further comparison of weights of scrap per unit with seed, and led to the interesting observation that there is a reverse correlation of animal bones and seeds. Further, comparing the abundance of bone with amount of soil from individual findspots made it possible to judge whether the quantity of bone found was related to the amount of earth moved or to a localized scarcity or abundance which would need some other explanation. When the areas excavated as units differ markedly in volume, this type of information is especially helpful.

As a result of our efforts over the three seasons discussed, a number of points can be made about dealing with the total bone recovered. First some comments on techniques: It would be better to measure quantities of scrap by weight than by count and by bulk as we did. To have a person at the excavation, not only to observe but also to supervise excavation of delicate pieces and to mend broken bone, saves much time in the laboratory. A minimum of handling of the bone not only saves time but also reduces breakage; to this end, when the bone comes in from the excavation it should first all be washed; scrap can then be sorted, weighed, examined, and rebagged. While scrap should be processed daily on a lot by lot basis, identifiable bone is best dealt with element by element and when a rather large body of material has accumulated.

Consideration of the larger question of just what should be done with scrap depends on the site. At Çayönü, neither funds nor time permitted the kind of total excavation carried on over a period of many years, which has been possible elsewhere. Our 5% exposure, though something of a record for excavations in the late prehistoric horizons in the Middle East, is still just a sample, and time was of the essence. Further, a basic aim of the project was to complete the study of as much material as possible each season at the site. As a consequence of all of this, for the bones the best procedure was to define certain limited objectives which could be accomplished for the total material rather than dealing with a portion of the material in complete detail. With two people working on bone, some scrap analysis was possible as well as the identification, measuring, and recording of identifiable specimens. The results of our work show that field supervision, as well as sorting and weighing in the laboratory, can be carried out routinely without demanding too much time and produces useful information. For instance, location, abundance, and type of fragment contribute to an understanding of what was going on in what areas. All possible comparisons have not been made, but we do know, for instance, that we have found no dump areas and no associated parts of carcasses, that there are evidences of refill in some places, that a total abundance of bone apparently correlated with domestication in the later subphases as well as with a decline in seed, that an MNI count of Bos in the early subphases is supported by an abundance of large fragments, and that scrap from a hunter's camp differs sharply from that from a site such as ours. I do not think we have exhausted the possibilities of these kinds of comparisons and feel that quantitative and qualitative sorts of scrap as discussed above are of value.

A more detailed scrutiny of all fragments, as that undertaken in our studies of modified bone, is too time consuming for an excavation such as Çayönü. Carried out on samples of the bone, it produces interesting results. The same is true of efforts to fine screen or float for small pieces of bone; to try to do this for all soil would slow the work down more than results justify.

My conclusions on reviewing the efforts of all three seasons are that it is not worthwhile saving scrap unless there is a real probability that it will be studied. If any is saved, it should all be saved, and a realistic plan for dealing with scrap needs to be made at the beginning of an excavation taking into consideration the time and funds available as well as the principal focus of the project.

Note

1. The Çayönü excavations were a joint venture of the Prehistory Section of Istanbul University and of the Oriental Institute of the University of Chicago, substantially supported by National Science Foundation grants GS-50, GS-1986 and GS-30365, cf. Proc. Nat. Acad. Sci. USA 71: 568-72 (1974).

References

Lawrence, B. and C.L. Wright
   1968   "Preliminary report on the mammal remains of the 1968 excavations at Çayönü," unpublished manuscript.

Sadek-Kooros, H.
   1972   "Primitive bone fracturing: a method of research," *American Antiquity*, vol. 37, pp. 369-382.
   1975   "Intentional fracturing of bone: description of criteria," in A.T. Clason, editor, *Archaeozoological Studies*, pp. 139-150. Amsterdam, New York.

EFFECTS OF CONTEXT ON THE INTERPRETATION OF FAUNAL REMAINS: A CASE STUDY

Richard H. Meadow, Department of Anthropology
Peabody Museum, Harvard University

Introduction

A conception common to many archaeologists and and faunal analysts alike is that study of bone material begins, and indeed should begin, when a corpus of material is turned over by the excavator to the zooarchaeologist. The latter then proceeds to identify the material and to produce lists of species, measurements, and specimen counts within the broad, usually chronological, framework provided by the archaeologist. An analyst who works in this fashion is not part of an excavation project and therefore often has little understanding of how the material was gathered, of the contexts from which it came, and sometimes even of the significance of the results for the interpretation of human prehistory. In the past, the men and women who undertook faunal identifications were trained primarily in zoology, paleontology, or anatomy and their interests lay in bridging the gap between Pleistocene and modern faunas. But even today among zooarchaeologists whose primary interest lies in what bones can tell us about past human exploitation and manipulation of animal populations, there is a curious lack of concern for the limitations of archaeologically derived faunal data. Now that quantitative analysis of animal bones has largely replaced a mere qualitative understanding of the presence and absence of species, it is imperative that investigators be aware of the possible sources of patterning, bias, and error in the faunal corpus.

Unlike lithics, ceramics, or metals, most archaeological bone was never a by-product or an end-product of artifact manufacture. Because of its nature as the unwanted debris of human consumption, most bone was consciously or unconsciously modified only in the course of exploiting the products which it accompanied. Furthermore, most bone was disposed of at that stage of the exploitation process when it was deemed no longer worthwhile bothering with. Thus archaeological bone can be expected only to reflect and not to represent practices of animal selection, food preparation, and consumption.

Following selection and slaughter, an animal is passed through a complex of preparation activities which include skinning, disarticulation and/or segmentation, distribution, and cooking. During some or all of these activities, bones may acquire butchering marks, blow marks, breaks, and changes in surface color and molecular structure. Furthermore, at any stage during the preparation process, the disposal of unwanted debris may take place. Since different procedures often will be carried out in different places, bones of an animal may become widely scattered. Disposal which occurs at the site of any preparation or consumption activity can be termed primary disposal (after Schiffer 1972, 1976 and Meadow 1976).

Not all bones will be disposed of by letting them drop in place. Many specimens, especially in areas which are inhabited continuously, may be swept up and deposited in a disposal area distant from the activity area in which they actually became waste. Such seconday disposal procedures may serve to mix the results of individual activities and to sort the bone material according to size or durability. A good example of what may be a secondary deposit is the fifty-centimeter pile of bones found in an alleyway between two structures of Ali Kosh Zone B_2 (Hole et al. 1969, fig. 10). Following the bone disposal activities which are attendant upon preparation and consumption, any number of processes may significantly alter the nature of the faunal corpus. Scavengers such as rodents or carnivores may drag off, modify, or completely destroy large numbers of specimens. Exposure to the elements and to everyday life in a busy site will have a winnowing effect on the corpus, as will the physiochemical environment of deposition. Natural processes of erosion may completely destroy or or thoroughly mix part of the remains. Man may redeposit bone by digging pits, wells, or foundation trenches, by building walls, by leveling off or building up his site of occupation. Such spatial and temporal disturbances which frequently occurred on sites occupied for long periods of time, is here termed tertiary deposition.

Upon the point of excavation, then, any site contains within it a sample of the faunal materials which have resulted from the consumption and disposal activities of the former inhabitants and which have been more or less modified by later agencies. The degree to which it will be possible to recover past patterns of activity is directly dependent upon the degree to which it is possible to separate out comparable units of analysis and to discover the effects of non-random bone destruction within these units. Ideally, each step back through the chain of depositional, disposal, and modification processes requires control of factors which may have altered the nature of the faunal corpus during the preceding steps. Thus it is necessary to understand the effects of postdepositional changes on different units before it is possible to fully define the character of disposal activities. To understand the nature of the preparation complex, one needs to control for patterns created or destroyed during disposal and deposition. Finally, in order to derive accurate information on animal husbandry practices or on faunal exploitation, it is necessary to consider all post-mortem activities.

The Case of Hajji Firuz

The site of Hajji Firuz provides a good example of the effects of disposal practices on the interpretation of faunal remains (Meadow 1975).[1] Located in northwestern Iran in the Solduz Valley southwest of Lake Rezeiyeh, this "Neolithic" village site (5500-5000 C-14 years B.C.) was partly excavated in 1968 (Voigt 1976). Although very few bones were recovered overall (466 or c. 4 per cubic meter of deposit excavated), analysis of the faunal material in relationship to the plans of each of the four phases defined by the excavator shows that most bone came from contexts within or immediately around structures. The 189 mammal bones from 19 contexts inside structures were recovered from floor deposits; their origin is either primary or secondary disposal activities. Two interior disposal areas containing 62 and 54 specimens are especially noteworthy. By way of contrast, 151 mammal bones from trash areas outside of structures came from 41 contexts, with the single largest concentration being 11 specimens. Most of this exterior bone came from alleyways or from close to the sides of buildings and probably derived from secondary disposal activities. Large open areas were free from bone or contained only scattered fragments. An additional 126 specimens were recovered from 19 units but either the deposits could not be assigned to a phase or the type of context was not determinable. Because the volumes of fill removed from each excavation unit were not recorded, it is not possible to compare directly the densities of bone in the different areas of the site or at different periods of time. If we assume, however, that the available corpus of bones from Hajji Firuz is either a complete or an unbiased sample of the areas excavated, it is possible, by using bone counts, to attempt to determine the existence of patterns which may affect our interpretation of these data.

For purposes of this study, a subset of the faunal remains from Hajji Firuz was selected and grouped according to criteria of time, context type, and taxon. Since sheep/goat (*Ovis*/*Capra*) and pig (*Sus*) together contributed by far the largest number of bone finds to the deposits (319 of 340 identified specimens), investigation of their relative distributions by phase and context type seemed most worthwhile. Of particular interest was to determine whether the nature of an area excavated might affect the relative abundance of the remains of the two taxa eventually identified by the analyst. Since the bones of pig tend to be larger and more massive than those of the caprines, there was some expectation that differences in preservation might be discernible. As for context type, a dichotomy - interior/exterior - was chosen for a number of reasons. The small number of bones from most excavation units necessitated grouping the units. Furthermore, there is good reason to believe that conditions for preservation inside and outside of closed buildings might differ significantly. Material inside is protected from the elements and from dogs, the latter perhaps the single most important cause of bone destruction in Middle Eastern sites. Flannery notes that the scarcity of bone in sixth millennium Tepe Sabz is linked with the appearance there of the domestic dog (Hole et al. 1969, p. 314) Dog remains have been identified from Hajji Firuz (Meadow 1975) where 19 of 25 specimens bearing the marks of carnivore teeth were recovered from contexts outside of structures.

Table 1 lists bone find counts element by element; these are summarized in Table 2. Sheep and goats outnumber swine in both interior and exterior contexts when the site is taken as a whole but pig bones are proportionately better represented in exterior loci. Proceeding phase by phase, however, the situation is not always so straight forward, a condition compounded in some cases by the small size of the sample. In an attempt to better characterize the situation, a number of summary statistics were calculated.

A common measure of the extent to which a sample of attributes deviates from the table that would be expected given independence is *chi-square* (X^2). As used here, with bone counts, this statistic is a summary measure of discrepancy, a descriptive statistic, and not a sampling distribution. In order to use chi-square properly for purposes of statistical inference (i.e., as a sampling distribution), it is necessary that the sample under consideration be a random sample of the population being investigated, a requirement which implies the provision that the presence of one item in the sample neither increase nor decrease the chance that another item will be present. The corpus of bones that we are dealing with is not a random sample; it is in fact the population of mammal bone present in the areas excavated. Only to the degree that this body of data can be considered as a sample of bones from the rest of the site or as a sample from the slaughtered population, can it be used for purposes of statistical inference. Furthermore, it is not always safe to assume, especially in the case of loci with many bones, that each bone comes from a separate animal. One could perhaps circumvent some of these difficulties by using Minimum Number of Individuals, but for reasons discussed in detail by Casteel (1977) and Grayson (1973, 1978, and most compellingly, 1978a), the author has chosen not to use this measure.

The summary measure of discrepancy is calculated by the formula $X^2 = \xi[(O_{ij}-E_{ij})^2/E_{ij}]$, where O_{ij}

TABLE 1. Counts of identified Ovis/Capra and Sus bone finds from interior, exterior, and not assigned (n.a.) contexts of Hajji Firuz, Phases A - D, listed by body part

	OVIS/CAPRA											SUS									
	Interior					Exterior					n.a.	Interior					Exterior					n.a.
Phase	A	B	C	D	tot	A	B	C	D	tot	n.a.	A	B	C	D	tot	A	B	C	D	tot	n.a.
HORN C.	1	.	.	3	4	1	.	.	.	1	2	-	-	-	-	-	-	-	-	-	-	-
SKULL	1	.	.	4	5	.	.	.	.	0	0	.	.	1	.	1	3	.	4	.	7	3
MAXILLARY	.	.	.	.	0	.	1	.	.	1	2	.	1	.	.	1	3	.	1	1	5	6
MANDIBLE	3	1	2	.	6	3	3	1	.	7	7	4	.	.	1	5	1	1	3	.	5	6
TEETH	.	1	.	.	1	4	1	.	.	5	10	1	.	1	3	5	2	1	1	.	4	2
VERTEBRAE	4	.	.	4	8	.	1	1	.	2	8	3	1	.	.	4	.	.	.	.	0	0
RIBS	.	.	.	.	0	.	2	1	1	4	1	.	.	1	.	1	.	1	.	.	1	0
SCAPULA	1	.	.	.	1	1	2	.	2	5	3	1	1	.	.	2	.	2	3	.	5	0
pr.HUMERUS	2	.	.	1	3	.	.	.	.	0	0	.	.	.	.	0	.	.	.	.	0	0
sh.HUMERUS	2	.	.	1	3	1	.	.	1	2	1	.	.	.	.	0	.	.	.	.	0	1
ds.HUMERUS	4	.	.	2	6	1	2	.	.	3	1	1	.	.	.	1	1	1	.	.	2	1
pr. RADIUS	3*	.	.	1	4*	2*	.	.	.	2*	3*	.	1*	1*	.	2**	.	.	1*	.	1*	1
sh. RADIUS	.	.	1	1	2	3	1	.	.	4	2	.	.	.	.	0	.	.	.	.	0	0
ds. RADIUS	1*	.	.	.	1*	.	.	.	.	0	0	.	1*	1*	.	2**	.	.	1*	.	1*	1
pr. ULNA	.	.	.	.	0	1	.	1	.	2	2*	.	.	.	.	0	.	.	2*	.	2*	1
sh. ULNA	2	1	.	.	3	2*	.	.	.	2*	0	.	.	.	.	0	.	.	.	.	0	2
ds. ULNA	.	.	.	.	0	.	.	.	.	0	0	.	.	.	.	0	.	.	1*	.	1*	0
CARPAL	.	.	.	2	2	.	.	.	.	0	0	1	.	.	.	1	.	.	.	.	0	0
pr. METAC.	1*	1	.	.	2*	.	1	.	.	1	2	2**	.	.	.	2**	.	.	.	.	0	1*
sh. METAC.	.	.	.	1	1	1	1	.	.	2	3	.	.	.	.	0	.	.	.	.	0	0
ds. METAC.	1*	.	.	.	1*	.	.	.	.	0	0	3**	.	.	.	3**	.	.	.	.	0	1*
PELVIS	.	.	.	1	1	1	1	.	.	2	1	.	.	.	.	0	1	1	1	.	3	3
pr. FEMUR	2	.	1	.	3	.	1	1	1	3	0	.	.	.	.	0	.	.	.	.	0	0
sh. FEMUR	4	.	1	1	6	3	2	1	.	6	0	.	.	1	.	1	.	.	1	.	1	1
ds. FEMUR	3	.	1	3	7	.	.	.	.	0	0	.	.	.	.	0	.	.	.	.	0	0
pr. TIBIA	2	.	.	.	2	.	.	.	.	0	0	.	.	.	.	0	.	.	.	.	0	0
sh. TIBIA	2	.	1	.	3	3	.	1	.	4	5	.	.	.	.	0	1	1	.	.	2	0
ds. TIBIA	.	.	1	3	4	.	1	.	.	1	2	.	.	.	.	0	.	.	1	.	1	1
TALUS	1	2	.	.	3	.	1	.	.	1	2	.	1	.	.	1	1	.	.	.	1	0
CALCANEUS	1	1	.	.	2	.	.	.	.	0	1	.	.	.	.	0	.	.	.	.	0	1
TARSAL	.	.	.	.	0	.	.	.	.	0	0	.	.	.	.	0	.	.	.	.	0	0
pr. METAT.	3	.	.	1*	4*	.	.	.	.	0	0	.	.	1*	.	1*	.	1*	.	.	1*	0
sh. METAT.	.	.	.	.	0	.	3	.	.	3	2	.	.	.	.	0	.	.	.	.	0	0
ds. METAT.	.	.	.	1*	1*	.	.	.	.	0	0	.	.	1*	.	1*	.	1*	.	.	1*	0
PHALANX 1	1	.	.	1	2	1	1	1	.	3	0	2	.	.	.	2	.	.	.	.	0	1
PHALANX 2	.	.	.	.	0	.	.	.	1	1	0	.	.	.	.	0	.	.	.	.	0	0
PHALANX 3	.	.	.	.	0	.	.	.	.	0	0	.	.	1	.	1	.	.	.	.	0	0
HUM/FEMUR	.	.	.	.	0	1	.	.	.	1	0	.	.	.	.	0	.	.	.	.	0	0
METAPODIAL	.	.	.	.	0	.	.	.	.	0	0	.	.	.	.	0	.	.	.	.	0	1
Totals*	43	7	8	30	__88__	28	25	8	6	__67__	__59__	16	5	6	4	__31__	14	9	17	1	__41__	__33__

* = part of same bone as another entry in table. To calculate actual number of specimens, subtract 0.5 for each star.

is the observed cell frequency and E_{ij} equals the column total times the row total divided by the grand total (N). Chi-square, however, is affected by the size of the sample used in its calculation. Therefore, when the master tables are divided into partial tables in order to view the effects of a third variable, it is necessary to use a statistical measure which is corrected for sample size. Dividing chi-square by number (N) yields *phi-square* (ϕ^2) which is the measure of deviation from expectation expressed on a per-unit basis. Like chi-square, phi-square is a summary of how much difference there is in a table as a whole between the observed and expected frequencies. Both phi-square and its square root *phi* (ϕ) are also commonly used measures of association; for dichotomous variables, the former may assume values between 0 (no association) and 1 (perfect association) and the latter can range between -1 and +1. (See Cowgill 1977.)

Table 5 contains the chi-square, phi-square, and phi values calculated for the counts of Table 2. For Phases A through D taken as a whole ('Totals'), the phi-square statistic suggests a rather weak effect of one variable on the other. Tables 3 and 4 suggest that for the same overall group, while most bone in each kind of context is sheep/goat, these taxa are relatively more plentiful in interior contexts. By the same token, by the nature of percentages, pig remains

TABLE 2. Summary counts of bone finds listed in Table 1

	Phase A			Phase B			Phase C			Phase D			Totals			n.a.	TOT
	IN	OUT	T	IN	OUT	T	IN	OUT	T	IN	OUT	T	IN	OUT	T		
O/C	43	28	71	7	25	32	8	8	16	30	6	36	88	67	155	59	214
SUS	16	14	30	5	9	14	6	17	23	4	1	5	31	41	72	33	105
Tot	59	42	101	12	34	46	14	25	39	34	7	41	119	108	227	92	319

TABLE 3. Percentages of taxa within context type, calculated from Table 2

	Phase A			Phase B			Phase C			Phase D			Totals			n.a.	TOT
	IN	OUT	T	IN	OUT	T	IN	OUT	T	IN	OUT	T	IN	OUT	T		
O/C	73	67	70	58	74	70	57	32	41	88	86	88	74	62	68	64	67
SUS	27	33	30	42	26	30	43	68	59	12	14	12	26	38	32	36	33
Tot	100	100	100	100	100	100	100	100	100	100	100	100	100	100	100	100	100

TABLE 4. Percentages within taxon by context type, calculated from Table 2

	Phase A			Phase B			Phase C			Phase D			Totals			n.a./TOT
	IN	OUT	T	IN	OUT	T	IN	OUT	T	IN	OUT	T	IN	OUT	T	
O/C	61	39	100	22	78	100	50	50	100	83	17	100	57	43	100	28
SUS	53	47	100	36	64	100	26	74	100	80	20	100	43	57	100	31
Tot	58	42	100	26	74	100	36	64	100	83	17	100	52	48	100	29

TABLE 5. Summary statistics calculated from Table 2

Phase A	Phase B	Phase C	Phase D	Totals
$X^2 = 0.4538$	$X^2 = 0.9674$	$X^2 = 2.3448$	$X^2 = 0.0345$	$X^2 = 3.7097$
$\phi^2 = 0.0045$	$\phi^2 = 0.0210$	$\phi^2 = 0.0601$	$\phi^2 = 0.0008$	$\phi^2 = 0.0163$
$\phi = 0.0670$	$\phi = 0.1450$	$\phi = 0.2452$	$\phi = 0.0290$	$\phi = 0.1278$

are relatively more common in exterior areas. This pattern is common to all phases except B. The lowest phi-square is found in Phase D, a phenomenon perhaps brought on both by the small size of the sample from exterior contexts and by the fact that almost all of the bones from interior contexts come from one large deposit associated with an human secondary burial (28 O/C bones, 3 Sus). The next lowest phi-square is that of Phase A which has a much larger sample but also is characterized by one abnormal interior locus (30 O/C bones, 6 Sus). For Phases B and C, phi-square is larger than it is for the site taken as a whole and in both phases Sus bones outnumber those of O/C in exterior loci. Phase C is unique in that pig bones outnumber those of the caprines overall, while Phase B stands out in Tables 3 and 4 as the only temporal division where there are proportionately more O/C bones in exterior loci. This same pattern would hold for Phases A and D should the contributions of the two abnormal loci be discounted from the respective totals.

These results suggest, therefore, that while a weak relationship may exist between context type and taxon, there is a good deal of fluctuation between the individual phases. The effects on the statistics of the two abnormal loci are also noteworthy. Together they provide almost 30% of the 227 caprine and pig bones assignable to context and phase and 21% of the total identified bone from the site. Should the 58 O/C bones and 9 Sus bones be subtracted from the totals for all phases combined, the resulting phi-square would be much lower ($\phi^2 = 0.0017$). The existence of such units raises the thorny problem of inference beyond the level of the remains themselves. To what extent will the Hajji Firuz bones, when taken together, actually reflect the "mean economic activities carried out during the whole period of accumulation of a deposit" (Uerpmann 1973, p. 308)? If not the total corpus, would some other grouping provide a more 'accurate' estimate of the 'true' relative importance of different species? In order to better answer these queries, it is necessary to examine the bones element by element and taxon by taxon to determine whether differential preservation or deposition can be determined to have taken place.

The distribution of swine and caprine bones from Hajji Firuz by context, body part, phase, and taxon is presented in Table 1. The contrast in the overall distribution of elements of pig and caprine is striking. Combined skull, maxillary, and mandibular counts for sheep and goats are only 11% of cranial plus post-cranial totals (excluding horncores, vertebrae, ribs, and loose teeth). The analogous figure for pigs is 44%. Femora and humeri are the most common post-cranial caprine remains with their distally articulated neighbors (tibiae, radii, and ulnae) only a little less frequent. The numbers of post-cranial bones located proximally or distally to these limb complexes are markedly lower. For pig, the scapula is by far the most commonly represented element, with the other limb bones being present in less than half its frequency.

Similar to this difference between swine and caprine in the overall frequency of skeletal elements is a contrast in the way the elements are distributed within the site. Perhaps the most striking feature of the array of caprine elements in Table 1 is the differential preservation of bone in interior and exterior loci. Among the post-cranial remains, all articular ends characterized by large amounts of spongy bone were recovered from interior contexts. These portions of elements include: proximal humerus, proximal tibia, distal femur, trochanter major of femur, and distal radius. In addition, horncores, skull fragments, and vertebrae are markedly more common in interior loci than they are in exterior contexts: 17 interior to 4 exterior [4.25:1] versus 88 to 67 [1.3.:1] for the whole collection. Corresponding figures for long bones without articular surfaces or unfused diaphyses are 35 to 12 [2.92:1] but for bare long bone shafts, 18 to 23 [0.78:1].

The frequencies for pig bones arrayed in Table 1 show a pattern directly opposite in some respects to that for the caprine material. Skull and maxilla fragments are more common in exterior loci (12) than in interior contexts (2). Since the ratio for the corpus as a whole is 41 exterior to 31 interior, this distribution also suggests an association between context and element type. In addition, spongy articular ends with the exception of the distal radius (and this always part of a whole bone), are completely missing from the swine sample. Long bones with articular surfaces are equally distributed between context type (7:7), while long bone shafts come exclusively from exterior areas (3 specimens).

The above patterns can be explained by postulating the existence of secondary disposal procedures and differential preservation. The relative protection from weathering and from dogs given to bones deposited within structures has already been suggested. Burial of interior remains during processes of rebuilding and reflooring would also tend to insure their preservation. Given these conditions, a more complete range of elements would be expected from interior loci. This kind of recovery does indeed occur with the caprine remains but not with the pig bones. In exterior loci, survival of thicker and larger bone fragments would be favored. Pig bone is more massive than caprine bone on the whole, and this perhaps is one reason why there is relatively more pig in exterior loci at Hajji Firuz. Pig skull bone, in particular, is thick and stands up well to abuse. The presence of proportionately four times as manu swine cranial fragments as sheep/goat examples in exterior contexts has already been noted.

Another reason for the relatively great amount of pig bone in exterior loci and also an explanation for the lack of pig bone with spongy articular surfaces in interior contexts may lie in the area of disposal practices. Briefly stated, the hypothesis is that since swine bone tends to be larger than the caprine equivalent, it therefore would have a greater chance of being noticed, swept up, and heaved out of a structure. If bone weight can be taken as an indicator of bone size, then the evidence presented in Table 6 does support this hypothesis. The average bone from an exterior locus tends to be 1.5 times as heavy as one from an interior context. Ranges are very wide, however, so this can be viewed only as a tendency; standard deviations cannot be calculated for this material because it was weighed in groups. For all loci, the average pig bone weighs about twice the average caprine bone. For exterior loci, this ratio is only slightly different (1.9:1) while for interior contexts it is somewhat lower (1.5:1). These statistics suggest that while bones tend to be smaller in interior loci, they also tend more toward the same size, thus indicating that larger bones in general are selected against. An exception is the case of the burial in Phase D mentioned previously.[2] Sus bones in exterior loci average about 1.7 times the weight of those inside while caprine bones from outside are about 1.3 times the average weight of specimens from interior loci. These figures support the hypothesis that the lighter caprine bone is getting broken up more in exterior loci than is the swine bone.

Thus the explanation for differential occurrence of pig and caprine bone may lie not in the species difference per se but in the intrinsic difference in size and durability of the bones. The result of such sorting, however, would be under-representation of pig in interior loci and over-representation in exterior contexts.[3] This phenomenon has been suggested above using bone counts and is also evident using weights. Weighing bone is considered by some scholars as the best way to calculate relative meat contributions of different species (see Uerpmann 1973, p. 310). Table 6 shows that while O/C bones outweigh Sus bones 1.8:1 in interior loci, a slight reverse tendency is present in exterior loci - 0.9:1.

These considerations lead back again to the question of which grouping of bones from a site gives the best estimate of the relative importance of different species. One answer, defining 'best' as 'most useful,' is to choose all loci grouped together because such statistics are most comparable to figures from other sites. Interior loci provide the environment for the best preservation, but there is always the chance that the picture will be skewed by the extraordinary survival of remains of individual activities. Such remains may include the bones of the same animal which have not gone through the scattering process to which secondary disposal might tend to subject them. The opposite problem is also present, however: the removal from

TABLE 6. Bone weights (in grams) of mammal bones from Hajji Firuz.
A = totals, B = averages, C = counts and weights for
all specimens from assignable contexts

A	Phase A IN	OUT	Phase B IN	OUT	Phase C IN	OUT	Phase D IN	OUT	N.A.	TOT
O/C	290.5	229.4	28.5	242.5	101.0	107.0	157.5	25.0	674.0	1855.5
SUS	157.5	260.0	53.0	133.5	49.0	249.0	57.5	53.5	740.5	1753.5

B	IN	OUT	IN	OUT	IN	OUT	IN	OUT	N.A.	TOT
O/C	6.76	8.50	4.07	9.70	12.63	13.88	5.63	4.17	11.42	8.88
SUS	9.84	18.57	10.60	14.83	8.17	14.65	14.38	(1 spc.)	22.44	16.70

C	Phase A IN	OUT	Phase B IN	OUT	Phase C IN	OUT	Phase D IN	OUT	TOT IN	OUT
N	93	46	19	55	19	40	52	10	183	151
Wt(g)	702	494.5	132.5	512.5	182.0	339.0	245.5	210.0	1262.0	1556.0

average for interior = 6.90; average for exterior = 10.30

interior loci of larger bone and therefore of the bones of larger species. If one discounts possible single activity loci and assumes that the loss of large specimens from interior loci equals the loss of small specimens from exterior loci, some sort of proportion can be calculated which might be the 'best' as far as an accurate average is concerned.

The calculation of an exact figure for relative importance of species, while giving an air of scientific accuracy, is in fact both misleading and of little value. The primary purpose of this study has been to show that the nature of the areas excavated in a site can influence the results of a faunal study. Therefore, any comparison of faunal assemblages from different sites, to say nothing of different periods within the same site, will need to be more qualitative than quantitative in nature unless one is able to isolate and compare contextual units. Such an approach, taken to its ultimate conclusion, requires the faunal analyst to be as familiar with the archaeology of a site as he or she is with the bones. This degree of sophistication, while perhaps ideal, is neither practical nor really necessary so long as the analyst is willing to consider the possible effects of context and is able to test for them by being given the relevant information by an alerted archaeologist.

Notes

1. Some of the results of this study have already been anticipated in Meadow 1975. Since publication of that paper, some additional Hajji Firuz specimens have come into my hands and Mary Voigt has been able to characterize additional loci as interior or exterior. I wish to thank Thomas W. Beale, George Cowgill, and Barbara Lawrence for commenting on earlier drafts of this paper and Mary Voigt for answering, cheerfully, numerous requests for more information.

2. The weight of the antler and horn material found in this unusual context have been discounted from the totals for all statistics presented in Table 6. Retaining them in the calculations raises the average weight of O/C bone from interior contexts to 10.29 g. from 6.96 g. (total weight = 905.5 g. for 88 specimens as opposed to 577.5 g. for 83 specimens).

3. A marked difference in interior/exterior bone size and weight could also be produced by an excavator being more careful in the recovery of material from interior floors. Such selection probably did not occur at Hajji Firuz, however.

References

Casteel, R.W.
   1977   "Characterization of faunal assemblages and the minimum number of individuals determined from paired elements: continuing problems in archaeology," *Journal of Archaeological Science*, vol. 4, pp. 125-134.

Cowgill, G.L.
   1977   "The trouble with significance tests and what we can do about it," *American Antiquity*, vol. 42, pp. 350-368.

Grayson, D.K.
   1973   "On the methodology of faunal analysis," *American Antiquity*, vol. 38, pp. 432-439.
   1978   "Minimum numbers and sample size in vertebrate faunal analysis," *American Antiquity*, vol. 43, pp. 53-65.
   1978a "On the quantification of vertebrate archaeofaunas," *Advances in Archaeological Method and Theory*, vol. 1, in preparation.

Hole, F., K.V. Flannery, and J.A. Neely
   1969   *Prehistory and Human Ecology of the Deh Luran Plain.* Memoirs of the Museum of Anthropology, University of Michigan, no. 1. Ann Arbor

Meadow, R.H.
  1975 "Mammal remains from Hajji Firuz: a study in methodology," in A.T. Clason, editor, *Archaeozoological Studies,* pp. 265-283. Amsterdam/New York.
  1976 "Methodological concerns in zooarchaeology," in *Thèmes spécialisés, Prétirage,* Union Internationale des Sciences Préhistoriques et Protohistoriques, pp. 108-123. Nice.

Schiffer, M.B.
  1972 "Archaeological context and systemic context," *American Antiquity,* vol. 37, pp. 156-165.
  1976 *Behavioral Archeology.* New York.

Uerpmann, H.-P.
  1973 "Animal bone finds and economic archaeology: a critical study of osteo-archaeological method," *World Archaeology,* vol. 4, pp. 307-322.

Voigt, M.M.
  1976 "Hajji Firuz Tepe: an economic reconstruction of a sixth millennium community in western Iran." Dissertation. Univ. of Pennsylvania.

PART TWO

MEASUREMENTS

We [in Germany and in German-speaking Central Europe] are trained first morphologically, second morphologically, and third morphologically. We are trained mainly morphologically and ecologically. Now it is our problem that we have had in Germany some schools which never go out of that field and can work only zoologically and have not found out what the archaeologists want...Such a situation we had worldwide for a long time. These people were working only in lists and in such lists we find nothing...If you know the bones really [and] can see in every bone more than "it is the bone of a horse, it is a bone of a donkey, or it is a bone of a sheep or a goat," if you can see the muscle marks--this muscle mark is so big because it is a male and an old male and this muscle mark is smaller because it is a female and a younger one and so on-- if you know your material in this manner, you can work. If you do not, you cannot work [J. Boessneck, 9 May 1975].

The preceding comments by Boessneck, abstracted from remarks made during the symposium, make clear his belief in the critical importance of a strong anatomical foundation for all osteo-archaeological research with particular attention to the bio-morphological characteristics of the bones studied. We quote the comment here because another dimension of this same concern for morphology is dealt with in the two papers included in this part--papers which present a case for the importance of taking comprehensive and multiple measurements of osteoarchaeological remains.

Measurement as a form of documentation is traditional in Central Europe where the investigation of differences in size and proportions of domestic animals has been particularly emphasized. Many scholars who have not had occasion to work on Central European material are largely unfamiliar with the literature, most of it in German. We hope that the papers in this part will serve to fill an existing void by showing what has been and can be done using measurements and by providing an introduction to a vast literature in which a large quantity of measurement data and evaluation has been published. The examples included in the paper of Boessneck and von den Driesch come almost entirely from Europe because only there have enough measurements been taken in a standardized fashion to permit statistically meaningful answers to be made to the questions asked. Uerpmann's paper in this section and those of Compagnoni in Part Four mark the beginnings of similar attempts in the Middle East.

Measurements are most useful if there are a lot of them, and thus there arises the question of how best to accumulate and publish such material. This matter was debated during the symposium and is reflected in the papers. Boessneck and von den Driesch emphasize that measurements must be published in detail but state that "the complete publication of individual measurements is an expensive undertaking and therefore has its limits." Recognizing these limits but also recognizing that the availability of measurements bone by bone is of great utility, Uerpmann proposes the establishment of a computerized data bank facility which would collect, store, and distribute this information. The difficulty with such an approach is that it, too, can be expensive, and the potential for abuse is great particularly if the measurements themselves are separated from the qualifications and descriptions of the original analyst. All measurements are raw data which, to be interpreted properly, must be used for a well-defined purpose in association with other, often non-quantifiable, data—both osteological and archaeological. Of course, measurements distributed in any form can be used inappropriately. It is thus important that the reasons for taking particular measurements be discussed when they are presented and their significance be evaluated.

R.H.M.
M.A.Z.

THE SIGNIFICANCE OF MEASURING ANIMAL BONES FROM ARCHAEOLOGICAL SITES

Joachim Boessneck and *Angela von den Driesch*, Institut für Palaeoanatomie,
Domestikationsforschung und
Geschichte der Tiermedizin,
University of Munich

That the study of animal bones from archaeological sites is informative from a culture-historical point of view is presumably well known.[1] The recovery of faunal remains along with archaeological finds has, as a result, become a matter of course. Restricting the study of faunal remains, however, to determination of the species present, their relative frequencies and ages, and perhaps even beyond this to specialized questions such as butchering practices by no means exhausts the available possibilities. Investigations of those kinds remain incomplete in one major respect, for in Europe at least it is understood that to objectify results and to gain additional information, it is necessary to take measurements on bones from animals which, at death, were more or less mature.

The collecting of measurements is done for the following reasons: zoological-systematic, ecological, and culture-historical. The measurements taken can be used for the following purposes, each of which will be illustrated below with examples to show the value of such an approach for osteoarchaeological investigations:

1. differentiation of animal species;
2. determination of individual variation;
3. investigation of size changes in wild animals through time;
4. investigation of size changes in wild animals through space;
5. determination of sex and the study of sexual dimorphism;
6. differentiation of domestic animals from their wild forebears;
7. documentation of size changes in domestic animals through the course of pre- and early historic times;
8. investigation of proportions and differences in proportions.

For the purpose of differentiating bones of closely related species (1), differences in size and proportions are often of primary importance. To determine these differences it is advantageous, if not indispensable, to take measurements. Should these measurements then be published, they will serve to underscore the identification. Should understandings change, conclusions based on the measurements can be revised. A case in point is the identification from bone remains of equid species; distinguishing between horse, onager, and ass is a particularly difficult and multi-faceted problem. Sometimes one reads a faunal list and finds, without further documentation, an attribution to one of the three species. In the case of equid identifications in the Near East, often some doubt may remain concerning the accuracy of that determination. Although Coon (1951) reports horse bones from western Iran in ancient times, the first verifiable account of the true horse in the Near East in the fourth millennium B.C. was reported during the symposium in Dallas. These finds come from eastern Anatolia (Boessneck and von den Driesch 1976). Since until now the Near East has been considered within the range of distribution of the onager alone, it would have been particularly worthwhile to have documented each horse bone from western Iran with measurements and illustrations. Such documentation has been undertaken for the finds from eastern Anatolia and with the equid bones from Mureybit (Ducos 1970). Taxonomic attribution of the latter, however, still remains in doubt even after the publication of measurement data. Furthermore, in the case of remains from later times when the keeping of horse and ass in the Near East had taken on a significant cultural role, the distinction between horse and ass, in spite of the size differences, can still be a problem. Thus Herre and Röhrs (1958) express themselves very circumspectly in reporting the results of their investigations of the equid bones from Osmankayasi. They do, however, present valuable lists of measurements which indicate that the small equids are donkeys. The situation becomes even more difficult for periods when one must allow for the presence of mules--crosses between horse and ass which fall between these species in size together with the onager and small horses. Identification may, therefore, remain questionable in some cases (cf. Kolb 1972; Boessneck and von den Driesch 1975; Krauss 1975). But to conclude from this situation that one should not take measurements, we consider wrong. A simple and clear listing of measurements will subsequently make possible clarification of the problems. In addition, comprehensive and detailed documentation is usable from points of view other than those of the original investigator. Of the osteoarchaeological works compiled before World War II, only those which contain comparable measurements are used today.

In order to define the individual variation in size and proportions (2) within a single population, extensive collections are necessary. For example, on the basis of quite a large series of red deer bones from one set of sites in Central Europe (e.g., Jéquier 1963; Blome 1968; W. Förster 1974; Scheck 1974/77), the amount of variation can be worked out. Such basic information is a prerequisite for the evaluation of small faunal samples as well as constituting the general foundation for all the types of investigations listed above (1-8).

In the case of domestication, total variation increases in almost every respect including animal size and proportions, features which can only be documented by macro-osteological methods. The extraordinary increase in variability is particularly clear when one considers the wide range evident between the breeds of dogs and compares it with the much narrower span which is to be observed for the dog's ancestor, the wolf. Such a variety of forms and races, however, together with the phenomenon of parallelism under domestication, are primarily concerns of zoological and domestic animal studies (cf. Herre and Röhrs 1973) and are not particularly relevant from an osteoarchaeological point of view. Under conditions of simple animal keeping without extreme selection, the form which is best fitted to the environment becomes the standard form. Variation remains narrow. Sexual dimorphism becomes reduced when compared with the wild forms (see below). Therefore, in the case of osteoarchaeological investigations, we have only to concern ourselves with the central part of the range of variation and with one size change (see below). Establishing these along with the origins of particular forms (cf., for example, Boessneck 1975b) are tasks of basic significance in questions relating to animal breeding.

Table 1 gives an idea of the variation in withers height among populations of pre- and early historic domestic animals. The data which was gathered together to ultimately be expressed in this summary comes from large bodies of faunal material such as those from the Celtic oppidum of Manching (Boessneck et al. 1971) and the moor sacrifice site of Skedemosse on Öland (Boessneck, von den Driesch, and Gejvall 1968), both dating to the Iron Age. We have chosen to deal with withers heights, which are used to denote the size of modern breeds of domestic animals, because doing so gives a better idea of the size of the animals than do the bone lengths from which they are calculated (cf. von den Driesch and Boessneck 1974). When comparing statements on withers heights, however, one must never forget to pay careful attention to how such heights are estimated. Otherwise rather remarkable discrepancies can occur, such as the case where one can calculate a height difference of 25 cm between Hittite horses from Osmankayasi and an Old Egyptian horse while, in fact, the animals were nearly the same size, to the extent that the metapodial lengths are practically identical (Boessneck 1970).

TABLE 1

Variation in withers height (WH) in populations of pre- and early historic domestic animals

	WH for example:
Horse: ca. 25 cm	1.20-1.45 m
Cattle: ca. 25 (up to 30) cm	1.00-1.25 (1.30) m
females: ca. 20 cm	1.00-1.20 m
males: ca. 20 cm	1.05-1.25 m
castrates: additional minimum of 5 cm	up to 1.30 m
Sheep: ca. 20-25 cm	0.50-0.75 m
females: ca. 20 cm	0.50-0.70 m
males: ca. 20 cm	0.55-0.75 m
Goats: ca. 20-25 cm	0.55-0.75 (0.80) m
females: ca. 15-20 cm	0.55-0.70 (0.75) m
males: ca. 15-20 cm	0.60-0.75 (0.80) m
Pigs: ca. 20 cm	0.60-0.80 m

Of particular interest from a zoological point of view is the determination of body size in wild species and its change through time (3). Just because wild animals in modern (remnant) populations are a certain size does not permit conclusions to be drawn about their size in ancient times. Some species were smaller in prehistoric times in the same regions where they are larger today; an example is the badger in Denmark (Degerbøl 1933). Other animals were larger in the past than they are today. Size reduction seems to have occurred at the beginning of the Holocene (cf. Requate 1957 on the aurochs) and to have continued into later prehistoric, early historic, and even more recent times (cf. Boessneck 1958a, pp. 47 ff.; Pietschmann 1977 on the red deer). To precisely determine such trends requires much additional investigation; larger series from the various time periods must be measured. One reason for size reduction may be that animals were excluded from the best biotopes by human groups which took them over for their own settlements; Jordan (1975) discusses one remarkable example of this--the case of the red deer in Thessaly (see our Fig. 1 and Table 2). Since size reduction appears also as a sequel to domestication (see below), it can be significant other than for purely zoological purposes to document measurements. Jarman (1972), for example, could develop his thesis in this way.

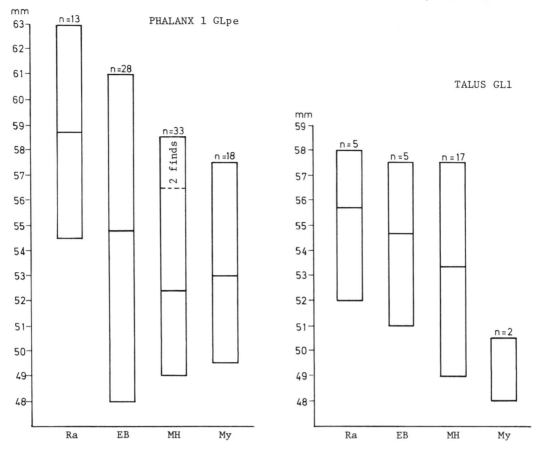

FIGURE 1. *Decrease over time in the size of red deer from the finds of Magula Pevkakia, Thessaly (after Jordan 1975, Diagram 4).*
GLpe = Greatest Length of peripheral half
GL1 = Greatest Length of lateral side

TABLE ?	Time	n	Variation	Mw	s	s%	$s_{\bar{x}}$
	Greatest Length of the Peripheral Half						
Measurements of the first phalanx of red deer from Magula Pevkakia, Thessaly (Jordan 1975, Table 32), to illustrate the size reduction from the Stone Age to the Bronze Age.	Ra	13	54.5 - 63.0	58.7	3.15	5.37	0.87
	EB	28	(48.0)- 61.0	54.8	2.96	5.41	0.56
	MH	33	49.0 - 58.5	52.4	2.41	4.60	0.42
	My	18	49.5 - 57.5	53.0	2.41	4.55	0.57
	Greatest Proximal Breadth						
	Ra	14	21.0 - 26.0	23.8	1.46	6.15	0.39
	EB	30	18.5 - 24.0	21.2	1.34	6.29	0.24
	MH	37	19.0 - 23.0	20.6	1.03	5.01	0.17
	My	20	19.0 - 23.0	21.1	1.29	6.11	0.29
	Smallest Breadth of Diaphysis						
Ra = Rachmani Period	Ra	10	16.5 - 21.0	19.3	1.34	6.94	0.42
EB = Early Bronze Age	EB	31	14.5 - 19.0	16.8	1.10	6.56	0.20
MH = Middle Helladic	MH	38	15.0 - 17.5	16.3	0.77	4.74	0.13
My = Mycenaean	My	19	15.0 - 19.0	16.6	1.24	7.49	0.28
	Greatest Distal Breadth						
	Ra	11	19.5 - 24.0	22.7	1.46	6.56	0.44
	EB	35	17.5 - 23.0	19.8	1.23	6.21	0.22
	MH	36	18.0 - 21.5	19.5	1.10	5.64	0.18
	My	19	18.0 - 21.5	19.8	1.23	6.21	0.28

Comparison of measurements gathered from a wide area may permit identification of regional size variations (4). In the case of red deer, the gross features of a size gradient from east to west has been known for a long time (e.g., Beninde 1937; Reichstein 1969a, 1969b). Finds from archaeological excavations, however, are particularly useful for increasing our knowledge on this score (e.g., Pietschmann 1977; see also our Fig. 2). Measurements of weasel bones from Anatolia (Boessneck 1974), when compared with data from other regions, show that the animals were distinctly and considerably different in size; this is a discovery of zoological and systematic significance and it was derived from osteoarchaeological materials.[2]

In order to correctly ascertain variation in size and proportions, it is indispensable to know the amount of sexual dimorphism present in a given species (5). The strongly marked size difference between males and females, particularly in ruminants but also in some carnivores, often makes possible the determination of sex from the size of certain bones alone, especially when the availability of large series permits the establishment of the complete range of variation. For examples of sexual dimorphism in deer see Figure 3 and Godynicki (1965); for bison see Koch (1932); for aurochs see Figure 4 and Degerbøl and Fredskild (1970); for other ruminants see Bosold (1966/1968). For carnivores, see Degerbøl (1933) and Eibl (1974) for the marten; Reichstein (1957) and

FIGURE 2. *Size comparison between prehistoric red deer bones from the Iberian Peninsula and from Central Europe (after von den Driesch 1972, Diagram 21 with additions).*

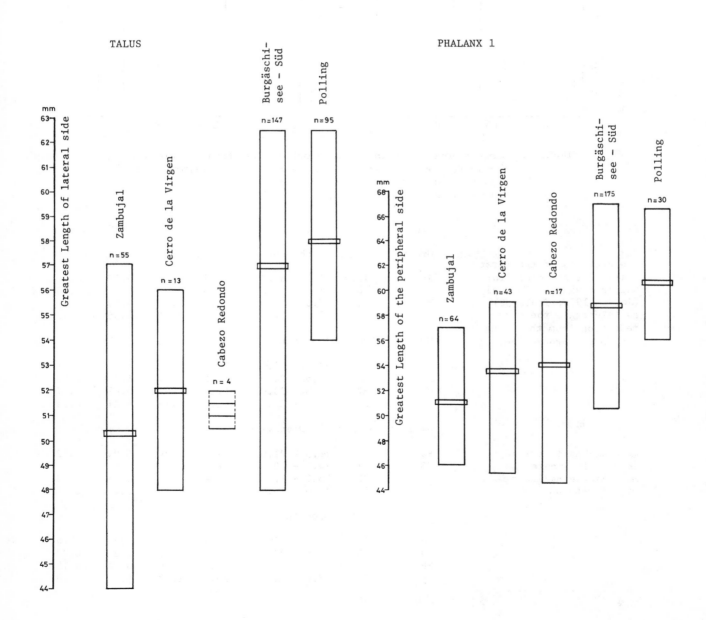

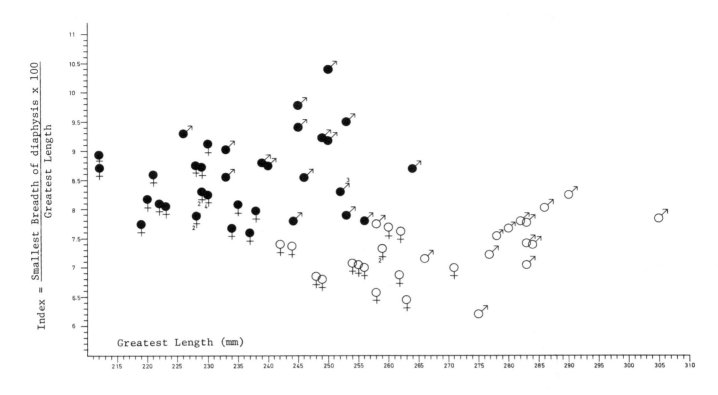

FIGURE 3. *Sexual dimorphism in metacarpals (●) and metatarsals (○) of red deer, based on recent material of known sex from Central Europe (after Bosold 1966/1968, Diagram 1).*

Boessneck (1974) for the weasel. Examples of determination of sex from the size of bones may be found in Geringer (1967: tables for red deer and elk); von den Driesch (1972: tables for red deer and ibex); and Boessneck and von den Driesch (1975, Table 32 for red deer). Of interest for culture history is whether, in the course of hunting, a selection according to sex was made by the hunters. For example, in the case of deer were more males than females singled out? To be sure, the interpretation of such findings is a rather new problem (e.g., Jéquier 1963, pp. 102 ff.; Blome 1968, pp. 21 ff.).

For domestic animals, it is important to know in what proportion males and females were kept so that one can draw conclusions about husbandry practices. In the case of cattle, one can use features of the metapodials in addition to other criteria to determine sex. Although the authors make sexual distinctions chiefly through visual evaluation of the bones--a procedure which permits the nonmeasurable specimens to be taken into consideration--the compilation of larger series of measurements is a worthwhile supplement for the demonstration of sexual dimorphism. The results are not usually so striking as those which are displayed in Figure 5. Sexual dimorphism decreases with reduction in body size (e.g., Nobis 1954), and with the practice of castration sex determination becomes more difficult. Even given these difficulties one should not forego bone measurements of small series just because these, when used alone, can be interpreted only with difficulty. Eventually measurements of material from the same period and region but from different sites can be brought together, as in Figures 6 and 7, and lead to informative interpretations.

Strongly marked sexual differences in size can cause considerable problems in distinguishing domestic animals from their wild forebears (6). These difficulties arise because in the initial phases of animal keeping, the sizes of bones from wild females overlap those from male domestic animals (e.g., Bökönyi 1962, diagrams). Fragments not able to be sexed are, therefore, not able to be characterized as wild or domestic. If, however, more and more measurements are collected, a judgment may be possible in many cases which are now problematical.

The recognition that measurements are useful for distinguishing wild and domestic animals (e.g., Fig. 4) is employed by Flannery (Hole, Flannery, and Neely 1969) in the cases of the length reduction of the M3 in pig (ibid., Fig. 129) and the size decrease in cattle bones (ibid., Fig. 126-127) as a result of domestication.

To determine the effects of culture and environment, it is necessary to work period by period in documenting the size of domestic animals from the various assemblages. Only in this way will it be possible to investigate size change through pre- and early historic times (7). In central Europe, size change has been examined for pig and cattle through measurements of the molar row, the

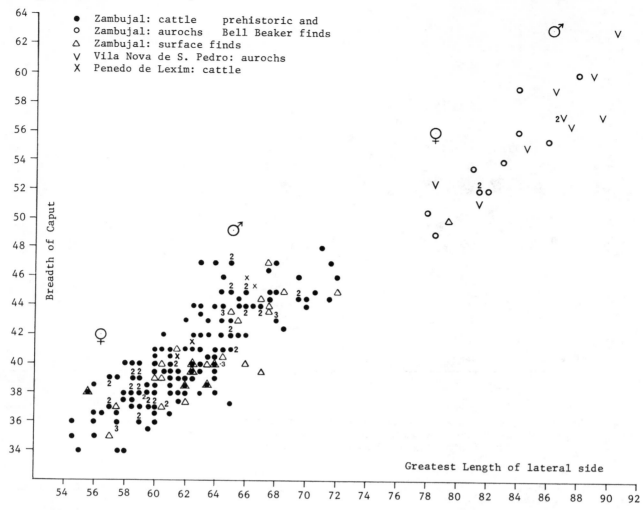

FIGURE 4. *Size differences between aurochs and Copper Age domestic cattle in Portugal and sexual dimorphism in the astragalus (after von den Driesch and Boessneck 1976, Diagram 2).*

FIGURE 5. *Sexual dimorphism in the metacarpus of prehistoric cattle from Zambujal, Portugal (after von den Driesch and Boessneck 1976, Diagram 1).*

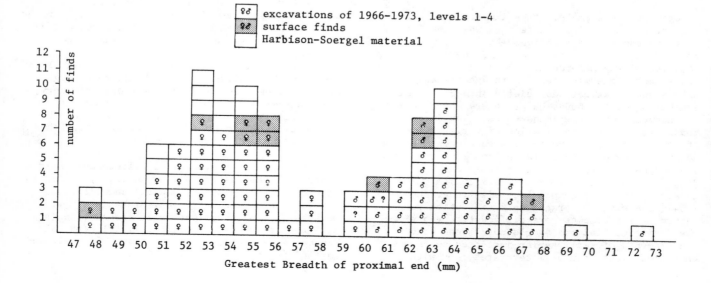

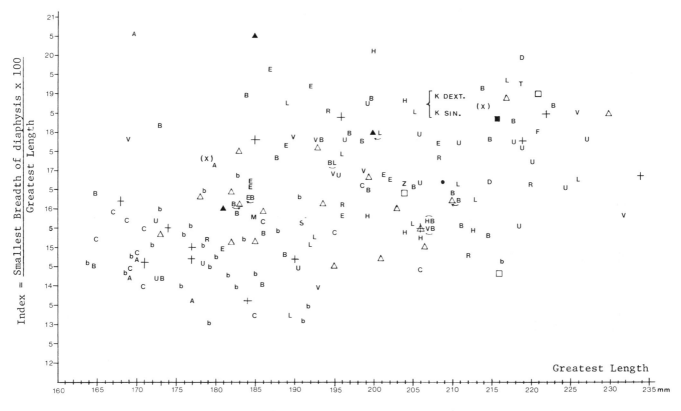

FIGURE 6. *Cattle metacarpals from Central Europe and Rumania Roman Period from Roman-occupied territory.*

FIGURE 7. *Cattle metacarpals from Central and Northern Europe Roman Period but from outside of Roman-occupied territory.*

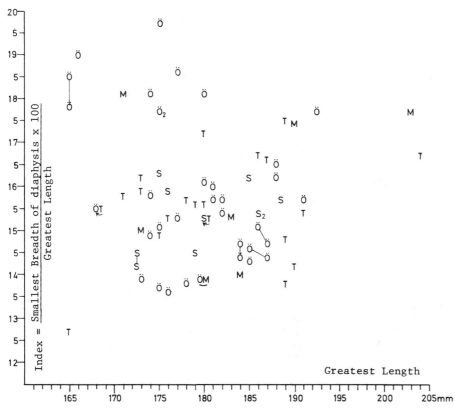

FIGURES 6 and 7. *Comparison of metacarpals of domestic cattle from Roman times in Roman-occupied and not Roman-occupied territory of Central Europe. The influence of the Romans can be recognized in the size increase in Roman-occupied territory (compare also Figures 8 and 9) as well as in the high proportion of steers and oxen which are to be found in the upper right of Figure 6. In Figure 7 the metacarpals of steers and of cows are similar in length; the broader metacarpal bones of steers are found above, the narrower bones of cows below. The dividing line lies somewhere about Index=17. Cut off from the others in the upper right are two metacarpals from oxen (after Boessneck et al. 1971, Diagrams XXXII and XXXIII, where also the find spot symbols are explained).*

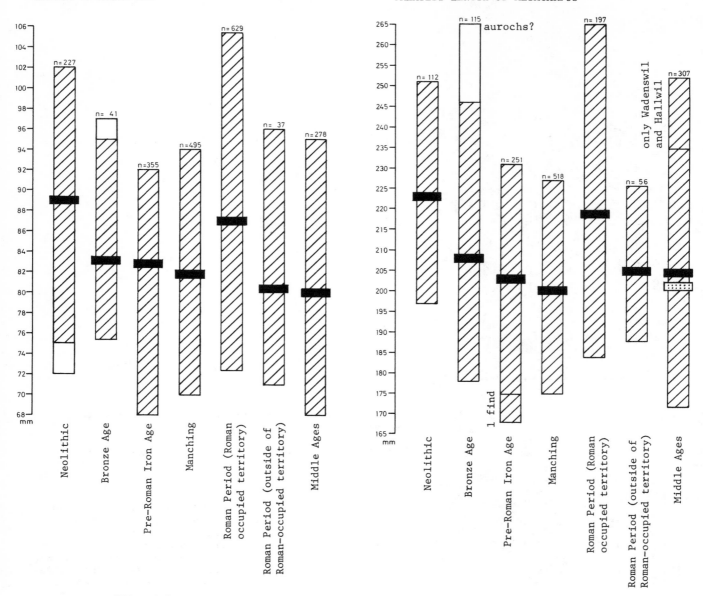

FIGURES 8 and 9. *Variation in the length of the molar row in mandibles and in the greatest length of metatarsus of domestic cattle in the major cultural periods. Material is from North and Central Europe as well as from Rumania (after Boessneck et al. 1971, Diagrams XIX and XLIX).*

M3, and the long bones (e.g., Boessneck 1958a, pp. 70 ff., 94 ff.; 1958b; Boessneck et al. 1971, pp. 56 ff.). Figures 8 and 9 show examples. These diagrams include single finds as well as larger series and when one knows that the measurements used were assembled merely because they were available (although value was placed in completeness, cf. Boessneck et al. 1971, pp. 56 ff.), then it is indeed remarkable with what uniformity both figures, along with others not included here (ibid., Diagrams XX, XXXVII, and XXXVIII), show the development. Furthermore, these bones, coming as they do from different parts of the body, are surely often not from the same animals and, in part, are also from different findspots. The same kind of size-change pattern can be followed in parts of Europe outside of that included in the figures: in England (Jewell 1962, 1963), in the Iberian Peninsula (von den Driesch 1972), and in Thessaly (Jordan 1975). This decline in animal size until the effects of Roman influence became apparent was clearly far reaching and independent of cultural group. The following explanation has been proposed by one of us (from Boessneck 1975a, pp. 178 ff.):

> It is first of all the expression of an accommodation to living conditions under which the smaller size brings advantages. Although conscious selection of smaller animals, especially of smaller males, in the earliest stages of domestication for the purpose of easing the human relationship with the cattle cannot be excluded, the primary reasons for the reduction in size must have been the shrinking of territory and altered selection pressure which, in a short period, reduced sexual dimorphism. Soon emphasis on numbers of animals as opposed to quality can be added as an additional factor, bringing as it did a situation where too many animals were kept for the land available and for the amount of fodder available in the winter. In various regions, especially in central Europe, can be added the fact that with population increase, the use of land for agriculture came at the expense of animal husbandry. Cattle became primarily dung and leather providers. This last development continued, for the most part, after Roman times and produced a size reduction up to the limits possible under simple conditions of animal keeping...[Fig. 8 and 9]... Where the Romans brought their cultural influence to bear strongly, there is a size increase evident in the animals. It is an expression of the first peak in selective animal breeding.

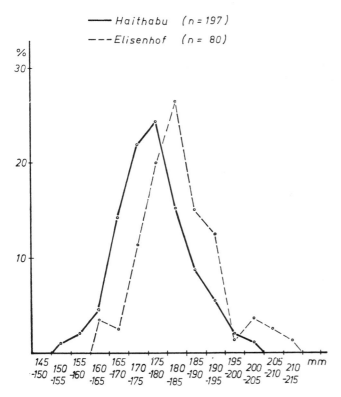

FIGURE 10. *Comparison of greatest length of metacarpus in cattle from an urban and a rural settlement in Schleswig-Holstein from the early Middle Ages (after Tiessen 1969, Illustration 4).*

These statements are based on measurements, a fact that shows clearly why such documentation should be carried out. Further investigations are essential in order to differentiate the various lines of development (e.g., Fig. 10 and Tiessen 1970; Reichstein 1973). Climate, through the environment, also influences the body size of animals. The pre-Bell Beaker and Bell Beaker pigs living in the rainier region under Atlantic influence on the northwestern edge of the Iberian Peninsula are larger than those pigs of the same periods living in the southeastern more semiarid lands (von den Driesch 1972, pp. 87 ff., 135 ff.).

It is of particular interest to collect series of measurements from the very diverse cultures and areas of the Near East. Surely animals in this varied area developed in very different ways and it will be a long time until we have more detailed information. As an example of how we present documentation for this area, our volume on the faunal remains from Korucutepe can be referred to (Boessneck and von den Driesch 1975). It is important to emphasize also that small groups of material must not be neglected (e.g., Boessneck 1973) for they too can lead to conclusions, particularly if used in an additive fashion (as in our Figs. 6 and 7). Should there be whole skeletons (e.g., Boessneck 1970, 1975b; Boessneck and von den Driesch, in press) or series of skeletons (e.g., Boessneck 1977a) available, measurements taken on these offer one the possibility of using them as a basis for interpretation should uncertainty over classification persist in the case of individual finds. Numerous such skeletons and series have been published from Central Europe, only a few of which, to introduce our method, are listed here: Boessneck and von den Driesch (1967): a wolf skeleton and four horse skeletons; Ekman (1972): an aurochs skeleton; Boessneck and Stork (1973): a pig skeleton; Stork and Boessneck (1975): fourteen horse skeletons.

For lands with "civilizations" where the animal world and life with animals has been portrayed in pictorial representations or in writing, bone measurements can serve to underpin conceptions of animal size and proportions. The combination of both methods of investigation—pictorial and documentary history with osteometry—is very important. For example, it is questionable whether the famous Old Egyptian greyhound, the tešem, ever, in fact, reached an extreme greyhound form. The representations on murals (e.g., Boessneck 1953, Fig. 23) are stylized. Skulls and skeletons of Old Egyptian dogs have to date never shown themselves to be particularly greyhoundlike (see Lortet and Gaillard 1903; Hauck 1941; Boessneck 1975b) but these come only rarely from the period of peak popularity of the tešem, the Old Kingdom.

Knowledge of the size of the various domestic animals is essential for drawing conclusions about paleoeconomies. If, for example, cattle are big and sheep are small, as was the case in Central Europe during the early Neolithic, then more than twenty sheep would be required to equal the weight of one cow (cf. Clason 1971, Table 8; 1973, Table 1). If, on the other hand, the cattle are hardly more than half as heavy but the sheep are rather larger than they were in the European Neolithic, as was the case in the pre-Roman Iron Age, then not even ten sheep will equal the weight of one cow (cf. Boessneck et al. 1971, p. 9; Boessneck and Krauss 1973, p. 124; Piehler 1976).

The investigation of proportions and differences in proportions (8) formed the basis for the discussion (above) of sexual differentiation using cattle metapodials. Another example might be distinguishing between the fore and hind phalanges of equids. Here measurements and index calculations offer extra aids to the morphological criteria which can be used without measurements (U. Förster 1960, diagrams). Finally, measurements of bones reflect not only animal size but also proportions and thus permit conclusions to be drawn about the environment.

So far all examples chosen have been from mammals. But bird bones also should be measured. Not only does this data serve to substantiate identification but little by little, basic information is accumulated which is valuable for ornithology (cf. Figs. 11 and 12). Difficulties in deciding whether one is dealing with domestic geese and ducks or with their wild forebears can be overcome with measurements. In the case of the chicken, measurements permit us to see the same special development in Roman times that occurred in the case of the domestic mammals (e.g., Schweizer 1961; Boessneck et al. 1971, pp. 92 ff.; Thesing 1977).

When all is said and done, it is clear that animal bones must be examined with as much care as other archaeological finds. And it is the responsibility of the analyst to bear in mind all possible points of view including those which approach zoological and culture-historical questions by examining significant measurements. Excavations are expensive. Excavated material is valuable. How much of this valuable material has already been lost as a result of incomplete documentation and insufficient interpretation?

To the questions of which measurements should be taken and how, von den Driesch in her guide gives the following answer (1976, p. 3):

> The decision on which skeletal parts to measure and which measurements should be taken must in the end be made by each researcher for himself. The choice of measurement depends on the value of a find and on the aims of the research. No norm can be set for the specialist. However if we hope, from the point of view of universal validity, to achieve comparable results, the use of the _same_ measurements is of the greatest importance in original research on site refuse.

The tradition of taking measurements reaches back to the beginnings of osteoarchaeological research —to L. Rütimeyer, Th. Studer, and A. Nehring. After World War II, the process of standardization carried forward an earlier tradition, that of Hilzheimer (e.g., 1920, 1924, 1941) and the Zurich school started by Hescheler with Wettstein (1924), Kuhn (e.g., 1932, 1935) and Rüeger (e.g., Hescheler and Rüeger 1942). Standardization in fact came about more or less of itself during that time since the same elements with the same

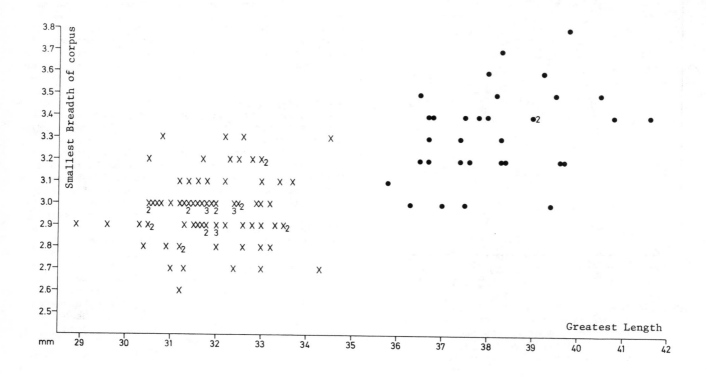

FIGURE 11. *Separation of ptarmigan species* Lagopus lagopus *(●) and* Lagopus mutus *(X) on the basis of the tarsometatarsus. From the Late Pleistocene levels in the Brillen cave near Blaubeuren in Württemberg (after Boessneck and von den Driesch 1973, Diagram IV).*

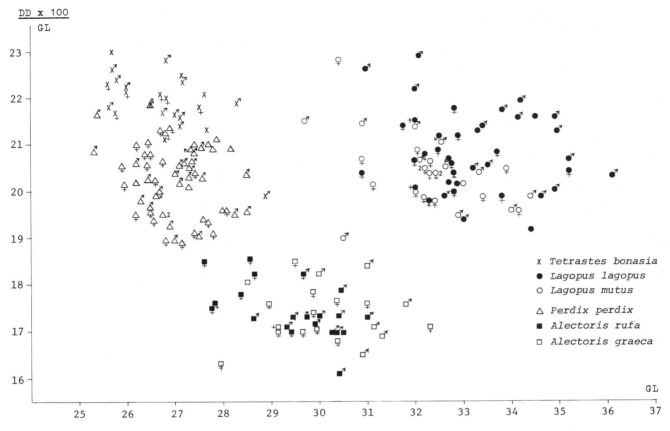

FIGURE 12. *Variation in size and proportions of the carpometacarpus of smaller galliformes of North and Central Europe (recent material of known sex, after Kraft 1972, Diagram 25).*
DD = greatest distal diagonal; GL = greatest length

measurement points were always being recovered from the settlement debris and one oriented oneself in the literature by what another had measured. Thus, in order to facilitate comparison, it seemed appropriate to take certain measurements.

Duerst (1926) did undertake to standardize measurements but his comprehensive work goes beyond the specialized concerns of osteoarchaeology. And because that work is all too rarely consulted, the confusion with respect to measurement points has not decreased with the result that published measurements are often not comparable. To help provide a solution to these problems, we have put together a special guide to bone measurement in osteoarchaeology which covers mammal and bird bones and includes those measurements which have been traditionally appropriate to take (von den Driesch 1976).

The complete publication of individual measurements is an expensive undertaking and therefore has its limits. but even though there may be compulsion to make summary statements, the detailed display of metrical results must not suffer in the publication of a corpus of material. One must take pains constantly to make sure that the material is set down in such form that it can be referred to in a complete sense, not only as underpinning for what is discussed in the publication, but also under circumstances not foreseen or not pursued by the investigator. The rarer, but still necessary, measurements which von den Driesch (1976) notes will become more valuable with time as more and more of them accumulate in the literature. In small bone samples they, together with the more commonly taken measurements, are indispensable to a complete evaluation of the material.

Working with bone measurements is, in the final analysis, a matter of experience. He who develops the expertise to a fine art will have success. But he who does not see the whole but loses himself in details risks drawing false conclusions. Grigson's comments (1969, pp. 288 ff.) on aurochs identified from Burgäschisee-Süd by Stampfli (1963) is a case in point. She did not recognize the particularly good conditions which existed in the case of the finds from Burgäschisee-Süd, perhaps because of the language barrier. For demonstration of her contentions, Grigson (1969, Fig. 6) chose a few measurements with short dimensions. She passed over the lengths of the completely preserved metacarpals. But it is just these lengths, when taken together with the breadth measurements, which permit no doubt at all either as to the identification as aurochs or as to sex. (For additional comments, see Boessneck 1977b.)

In conclusion, we hope that in this essay we have argued the case for measurements so convincingly by the selected examples that, in the future, there will no longer remain unused so much valuable, laboriously and expensively recovered faunal material.

Notes
1. Translated from the German by Richard H. Meadow.
2. Hole, Flannery, and Neely (1969, p. 317) illustrate an alleged mustelid mandible from Khuzistan. This find is either reproduced in a scale larger than that given or it is not a weasel. A statement of the length of the tooth row (cf. Gaffrey 1953, 1961) would have served here to further clarify the matter.

References

Beninde, J.
  1937 *Zur Naturgeschichte des Rothirsches.* Monographien der Wildsäugetiere, vol. 4, pp. 1-223. Leipzig.

Blome, W.
  1968 "Tierknochenfunde aus der spätneolithischen Station Polling." Dissertation, München.

Bökönyi, S.
  1962 "Zur Naturgeschichte des Ures in Ungarn und das Problem der Domestikation des Hausrindes," *Acta Archaeologica Akademiae Scientiarum Hungaricae,* vol. 14, pp. 175-214.

Boessneck, J.
  1953 *Die Haustiere in Altägypten.* Veröffentlichungen der Zoologischen Staatssammlung München, vol. 3, pp. 1-50. München.
  1958a *Zur Entwicklung vor- und frühgeschichtlicher Haus- und Wildtiere Bayerns im Rahmen der gleichzeitigen Tierwelt Mitteleuropas.* Studien an vor- und frühgeschichtlichen Tierresten Bayerns, vol. 2. München.
  1958b "Herkunft und Frühgeschichte unserer mitteleuropäischen landwirtschaftlichen Nutztiere," *Züchtungskunde,* vol. 30, pp. 289-296.
  1970 "Ein altägyptisches Pferdeskelett," *Mitteilungen des Deutschen Archäologischen Instituts Kairo,* vol. 26, pp. 43-47. Mainz.
  1973 "Tierknochenfunde vom Zendan-i Suleiman (7. Jahrhundert v. Christus)." *Archäologische Mitteilungen aus Iran,* n.F., vol. 6, pp. 95-111. Berlin.
  1974 "Eine vergleichende Dokumentation subfossiler Wieselfunde aus Anatolien," *Säugetierkundliche Mitteilungen,* vol. 22, pp. 304-313.
  1975a "Osteoarchäologie," *Ausgrabungen in Deutschland gefördert von der Deutschen Forschungsgemeinschaft 1950-1975,* Part 3, pp. 174-182. Mainz.
  1975b "Ein altägyptisches Hundeskelett aus der 11. Dynastie," *Mitteilungen des Deutschen Archäologischen Instituts Kairo,* vol. 31, pp. 7-13. Mainz.
  1977a "Die Hundeskelette von Isân Bahriyât (Isin) aus der Zeit um 1000 v. Chr.," in "Isin-Isân Bahriyât I: Die Ergebnisse der Ausgrabungen 1973-1974," *Bayerische Akademie der Wissenschaften (philosophisch-historische Klasse),* n.F., no. 79, pp. 97-133. München.
  1977b "Die Tierknochenfunde aus der Siedlung der Rössener Kultur Schöningen, Eichendorffstrasse, und die Probleme ihrer Ausdeutung," *Neue Ausgrabungen und Forschungen in Niedersachsen,* vol. 11, pp. 153-158. Hildesheim.

Boessneck, J. and A. von den Driesch
  1967 "Die Tierknochenfunde aus dem fränkischen Reihengräberfeld in Kleinlangheim, Landkreis Kitzingen," *Zeitschrift für Säugetierkunde,* vol. 32, pp. 193-215.
  1973 *Die jungpleistozänen Tierknochenfunde aus der Brillenhöhle.* Das Paläolithikum der Brillenhöhle bei Blaubeuren (Schwäbische Alb) II. Stuttgart.
  1975 "Tierknochenfunde vom Korucutepe bei Elâzığ in Ostanatolien (Fundmaterial der Grabungen 1968 und 1969)," in M.N. van Loon, editor, *Korucutepe I,* pp. 1-220. Amsterdam/New York.
  1976 "Pferde im 4./3. Jahrtausend v. Chr. in Ostanatolien," *Säugetierkundliche Mitteilungen,* vol. 24, pp. 81-87.
  in press "Die zoologische Dokumentation von 3 Pferdeskeletten und anderen Tierknochenfunden aus einem Kammergrab auf dem Norsun-Tepe, Ostanatolien," *Istanbuler Mitteilungen.*

Boessneck, J. and R. Krauss
  1973 "Die Tierwelt um Bastam/Nordwest-Azerbaidjan," *Archäologische Mitteilungen aus Iran,* n.F., vol. 6, pp. 113-133. Berlin.

Boessneck, J. and M. Stork
  1973 "Die Tierknochenfunde der Ausgrabungen 1959 auf der Wüstung Klein-Büddenstedt, Kreis Helmstedt," *Neue Ausgrabungen und Forschungen in Niedersachsen,* vol. 8, pp. 179-213. Hildesheim.

Boessneck, J.A., A. von den Driesch, and N.-G. Gejvall
  1968 *Knochenfunde von Säugetieren und vom Menschen.* The Archaeology of Skedemosse III. Stockholm.

Boessneck, J., A. von den Driesch, U. Meyer-Lemppenau, and E. Wechsler-von Ohlen
  1971 *Die Tierknochenfunde aus dem Oppidum von Manching.* Die Ausgrabungen in Manching 6. Wiesbaden.

Bosold, K.
  1966/68 "Geschlechts- und Gattungsunterschiede an Metapodien und Phalangen mitteleuropäischer Wildwiederkäuer." Dissertation. München. 1966; *Säugetierkundliche Mitteilungen,* vol. 16, pp. 93-153. München. 1968.

Clason, A.T.
1971 "The flint-mine workers of Spiennes and Rijckholt-St. Geertruid and their animals," *Helinium*, vol. 11, pp. 3-33.
1973 "Some aspects of stock-breeding and hunting in the period after the Bandceramic culture North of the Alps," in H. Matolcsi, editor, *Domestikationsforschung und Geschichte der Haustiere*, pp. 205-212. Budapest.

Coon, C.S.
1951 *Cave Explorations in Iran 1949*. Philadelphia.

Degerbøl, M.
1933 *Danmarks Pattedyr i Fortiden i Sammenligning med recent Former*. Videnskabelige Meddelelser fra Dansk naturhistorisk Forening, vol. 96, part 2.

Degerbøl, M. and B. Fredskild
1970 *The Urus (Bos primigenius Bojanus) and Neolithic Domesticated Cattle (Bos taurus domesticus Linné) in Denmark*. Det Kongelige Danske Videnskabernes Selskab Biologiske Skrifter, vol. 17, no. 1.

Driesch, A. von den
1972 *Osteoarchäologische Untersuchungen auf der Iberischen Halbinsel*. Studien über frühe Tierknochenfunde von der Iberischen Halbinsel, vol. 3, pp. 1-267. München.
1976 *A Guide to the Measurement of Animal Bones from Archaeological Sites*. Peabody Museum Bulletin 1. (Harvard University); also as "Das Vermessen von Tierknochen aus vor- und frühgeschichtlichen Siedlungen." Institut für Palaeoanatomie, München.

Driesch, A. von den and J. Boessneck
1974 "Kritische Anmerkungen zur Widerristhöhenberechnung aus Längenmassen vor- und frühgeschichtlicher Tierknochen," *Säugetierkundliche Mitteilungen*, vol. 22, pp. 325-348.
1976 "Die Tierknochenfunde vom Castro do Zambujal," *Studien über frühe Tierknochenfunde von der Iberischen Halbinsel*, vol. 5, pp. 4-129. München.

Ducos, P.
1970 "Les restes d'équides. Part IV, The Oriental Institute Excavation at Mureybit, Syria, Preliminary report on the 1965 Campaign," *Journal of Near Eastern Studies*, vol. 29, pp. 273-289.

Duerst, J.U.
1926 "Vergleichende Untersuchungsmethoden am Skelett bei Säugern," in *Handbuch der biologischen Arbeitsmethoden*, Abt. 7. Methoden der vergleichenden morphologischen Forschung, Heft 2, pp. 125-530. Berlin and Wien.

Eibl, F.
1974 "Die Tierknochenfunde aus der neolithischen Station Feldmeilen-Vorderfeld am Zürischsee, I: Die Nichtwiederkäuer." Dissertation. München.

Ekman, J.
1972 "The urus female (Bos primigenius Boj.) from Slagarp, Southern Sweden," *Zoologica Scripta*, vol. 1, pp. 203-205.

Förster, U.
1960 *Die Pferdephalangen aus dem keltischen Oppidum von Manching*. Studien an vor- und frühgeschichtlichen Tierresten Bayerns, vol. 8. München.

Förster, W.
1974 "Die Tierknochenfunde aus der neolithischen Station Feldmeilen-Vorderfeld am Zürichsee, II: Die Wiederkäuer." Dissertation. München.

Gaffrey, G.
1953 *Die Schädel der mitteleuropäischen Säugetiere*. Abhandlungen und Berichte aus dem Staatlichen Museum für Tierkunde, Forschungsinstitut Dresden, vol. 21. Leipzig.
1961 *Merkmale der Wildlebenden Säugetiere Mitteleuropas*. Leipzig.

Geringer, J.
1967 *Tierknochenfunde von der Heuneburg, einem frühkeltischen Herrensitz bei Hundersingen an der Donau (Grabungen 1959 und 1963), Die Paarhufer ohne die Bovini*. Dissertation. München; also in *Naturwissenschaftliche Untersuchungen zur Vor- und Frühgeschichte in Württemberg und Hohenzollern*, vol. 5. Stuttgart.

Godynicki, S.
1965 "Determination of deer height on the basis of metacarpal and metatarsal bones" (in Polish with English and Russian summaries), *Roczniki Wyzszej, Szkoly Rolniczej w Poznaniu*, vol. 25, pp. 39-51.

Grigson, C.
1969 "The uses and limitations of differences in absolute size in the distinction between the bones of aurochs (Bos primigenius) and domestic cattle (Bos taurus)," in P.J. Ucko and G.W. Dimbleby, editors, *The Domestication and Exploitation of Plants and Animals*, pp. 277-294. London.

Hauck, E.
1941 *Die Hunderassen im alten Ägypten*. Zeitschrift für Hundeforschung, n.F., vol. 16. Leipzig.

Herre, W. and M. Röhrs
1958 "Die Tierreste aus den Hethitergräbern von Osmankayasi," in K. Bittel et al., *Die Hethitischen Grabfunde von Osmankayasi. Bogazköy-Hattusa II*. Wissenschaftliche Veröffentlichung der Deutschen Orient-Gesellschaft, no. 71, pp. 60-80.
1973 *Haustiere zoologisch gesehen*. Stuttgart.

Hescheler, K. and J. Rüeger
1942 "Die Reste der Haustiere aus den neolithischen Pfahlbaudörfern Egolzwil 2 (Wauwilersee, Kt. Luzern) und Seemate-Gelfingen (Baldeggersee, Kt. Luzern)," *Vierteljahrsschrift der Naturforschenden Gesellschaft in Zürich*, vol. 87, pp. 383-486.

Hilzheimer, M.
1920 "Die Tierreste aus dem römischen Kastell Canstatt bei Stuttgart und anderen römischen Niederlassungen in Württemberg," *Landwirtschaftliche Jahrbücher*, vol. 55, pp. 293-336.
1924 "Die im Saalburgmuseum aufbewahrten Tierreste aus römischer Zeit," *Saalburg-Jahrbücher*, vol. 5, pp. 105-158.
1941 *Animal Remains from Tell Asmar*. Studies in Ancient Oriental Civilization, no. 20. Chicago.

Hole, F., K.V. Flannery, and J.A. Neely
1969 *Prehistory and Human Ecology of the Deh Luran Plain*. Memoirs of the Museum of Anthropology, University of Michigan, no. 1. Ann Arbor.

Jarman, M.R.
1972 "European deer economies and the advent of the neolithic," in E.S. Higgs, editor, *Papers in Economic Prehistory*, pp. 125-147.

Jéquier, J.-P.
1963 Contributions in J. Boessneck, J.-P. Jéquier, and H.R. Stampfli, *Seeberg Burgäschisee-Süd, Teil 3, Die Tierreste*. Acta Bernensia II. Bern.

Jewell, P.A.
1962 "Changes in size and type of cattle from prehistoric to mediaeval times in Britain," *Zeitschrift für Tierzüchtung und Züchtungsbiologie*, vol. 77, pp. 159-167.
1963 "Cattle from British Archaeological Sites," *Occasional Paper No. 18 of the Royal Anthropological Institute*, pp. 80-101. London.

Jordan, B.
1975 "Tierknochenfunde aus der Magula Pevkakia in Thessalien." Dissertation. München.

Koch, W.
1932 *Über Wachstums- und Altersveränderungen am Skelett des Wisents*. Beiträge zur Natur- und Kulturgeschichte Lithauens und angrenzender Gebiete. Abhandlung der mathematisch-naturwissenschaftlichen Akademie der Wissenschaften, Suppl.-Band 15. Abhandlung, pp. 553-678. München.

Kolb, R.
1972 "Die Tierknochenfunde vom Takht-i Suleiman in der Iranischen Provinz Aserbeidschan (Fundmaterial der Grabung 1969)." Dissertation. München.

Kraft, E.
1972 "Vergleichend morphologische Untersuchungen an Einzelknochen nord- und mitteleuropäischer kleinerer Hühnervögel." Dissertation. München.

Krauss, R.
1975 "Tierknochenfunde aus Bastam in Nordwest-Azerbaidjan/Iran (Fundmaterial der Grabungen 1970 und 1972)." Dissertation. München.

Kuhn, E.
1932 "Beiträge zur Kenntnis der Säugetierfauna der Schweiz seit dem Neolithikum," *Revue Suisse de Zoologie*, vol. 39, pp. 531-768.

Lortet, A. and C. Gaillard
1903 "La faune momifiée de l'ancienne Egypte," *Archives du Muséum d'histoire naturelle de Lyon*, vol. 8, pp. 1-18.

Nobis, G.
1954 "Zur Kenntnis der ur- und frühgeschichtlichen Rinder Nord- und Mitteldeutschlands," *Zeitschrift für Tierzüchtung und Züchtungsbiologie*, vol. 63, pp. 155-194.

Piehler, W.
1976 "Die Knochenfunde aus dem spätrömischen Kastell Vemania." Dissertation. München.

Pietschmann, W.
1977 "Zur Grösse des Rothirsches (*Cervus elaphus* L.) in vor- und frühgeschichtlicher Zeit (Untersuchungen an Knochenfunden aus archäologischen Ausgrabungen)." Dissertation. München.

Reichstein, H.
1957 "Schädelvariabilität europäischer Mauswiesel (*Mustela nivalis* L.) und Hermeline (*Mustela erminea* L.) in Beziehung zu Verbreitung und Geschlecht," *Zeitschrift für Säugetierkunde*, vol. 22, pp. 151-182.
1969a "Untersuchungen von Geweihresten des Rothirsches (*Cervus elaphus* L.) aus der frühmittelalterlichen Siedlung Haithabu (Ausgrabung 1963-1964)," *Berichte über die Ausgrabungen in Haithabu*, Bericht 2, pp. 57-70. Neumünster.
1969b "Zur Frage der Herkunft der Rothirschgeweihe von Haithabu," in J. Boessneck, editor, *Archäologie und Biologie*. Forschungsberichte 15 der Deutschen Forschungsgemeinschaft, pp. 69-75. Wiesbaden.
1973 "Untersuchungen zur Variabilität frühgeschichtlicher Rinder Mitteleuropas," in H. Matolcsi, editor, *Domestikationsforschung und Geschichte der Haustiere*, pp. 325-340. Budapest.

Requate, H.
1957 "Zur Naturgeschichte des Ures (*Bos primigenius* Bojanus 1827) nach Schädel- und Skelettfunden in Schleswig-Holstein," *Zeitschrift für Tierzüchtung und Züchtungsbiologie*, vol. 70, pp. 297-338.

Scheck, K.
1974/77 "Die Tierknochen aus dem jungsteinzeitlichen Dorf Ehrenstein (Gemeinde Blaustein, Alb-Donau-Kreis)," Dissertation. München. 1974; also in *Forschung und Berichte zur Vor- und Frühgeschichte in Baden-Württemberg*, vol. 9, pp. 1-69. Stuttgart, 1977.

Schweizer, W.
1961 "Zur Frühgeschichte des Haushuhns in Mitteleuropa," *Studien an vor- und frühgeschichtlichen Tierresten Bayerns,* vol. 9. München.

Stampfli, H.R.
1963 Contributions in J. Boessneck, J.-P. Jéquier, and H.R. Stampfli, *Seeberg Burgäschisee-Süd, Teil 3, Die Tierreste.* Acta Bernensia II. Bern.

Stork, M. and J. Boessneck
1975 "Die Tierskelette aus dem awarischen Gräberfeld Wien-Liesing," *Mitteilungen der Anthropologischen Gesellschaft in Wien,* vol. 104, pp. 115-137. Wien.

Thesing, R.
1977 "Die Grössenentwicklung des Haushuhns in vor- und frühgeschichtlicher Zeit." Dissertation. München.

Tiessen, M.
1969 "Die Tierwelt einer städtischen und einer ländlichen Siedlung im frühmittelalterlichen Schleswig-Holstein," in J. Boessneck, editor, *Archäologie und Biologie.* Forschungsberichte 15 der Deutschen Forschungsgemeinschaft, pp. 148-156. Wiesbaden.
1970 "Die Tierknochenfunde von Haithabu und Elisenhof." Dissertation. Kiel.

Wettstein, E.
1924 "Die Tierreste aus dem Pfahlbau am Alpenquai in Zürich," *Vierteljahrschrift der Naturforschenden Gesellschaft in Zürich,* vol. 69, pp. 78-127.

METRICAL ANALYSIS OF FAUNAL REMAINS FROM THE MIDDLE EAST

Hans-Peter Uerpmann, Institut für Urgeschichte
University of Tübingen

Metrical analysis of faunal remains has a long tradition, particularly in the Central European school of zooarchaeology.[1] We cannot present here a complete survey of the history and development of this subject, but must note that metrical analysis of animal bones has been carried out by several generations of scholars for more than a hundred years. As a result, a relatively high degree of standardization of methods has been achieved. Since publication of the fundamental work of Duerst (1926) on measuring techniques, there has been general agreement on what to measure and how to measure it. But unfortunately this agreement has extended only to those countries in which the German language is usually read by scholars; thus northern and eastern Europe became incorporated into the tradition of metrical analysis and contributed to its development. The Anglo-American branch of zooarchaeology has never been closely connected to this tradition with the result that there still exists some misunderstanding of the techniques, as well as of the aims and possibilities, of metrical analysis.

We cannot deal here with measuring techniques in detail. Only two basic questions will be considered. The first is whether all of the many measurements that have been defined for the different bones are necessary, or whether their number could be reduced to one or two per bone, or even to some few representative measurements per animal. Since metrical analysis was carried to extremes by physical anthropology about fifty years ago, there is a tendency among scholars to reduce the number of measurements--a tendency that has been reinforced by a general reluctance to take measurements among scholars in the "humanities" (to which physical anthropology as well as zooarchaeology have close connections). A solid argument for a reduction in measurements could be made from statistical considerations, because many different measurements on one bone show high degrees of correlation to one another. Taking them all will therefore result in redundancy. Although it would thus be possible to characterize a complete bone by a few measurements, it is nevertheless important to take as many measurements as there are easily defined measuring points on a particular skeletal element.

The reason for this is that zooarchaeology normally does not deal with complete bones, but rather with fragments of all sizes. Defining as many measurements as feasible increases the chances that at least one measurement will be possible on a fragment. Thus the number of bones that can be included in metrical analysis is increased, with the result that statistical errors are reduced. Thus when dealing with fragments, taking apparently redundant measurements yields more complete information.

The second basic question of technique is whether the definition of a bone measurement can be made so precise that differences in the results obtained when different persons take this measurement can be avoided. In most cases it is not sufficient to define a measurement by its specification alone. A "distal width of the tibia" can be measured in very different ways giving different results. Even if it is called "maximum distal width of the tibia" the specification is not sufficient, because now the orientation of the measurement becomes important in that the greatest extension of the distal end may not be its "width." However, if measurements are defined by detailed descriptions (verbal and/or pictorial) comparable results will generally be produced. The most convincing evidence for this statement is the fact that in Central Europe--where a common tradition of bone-measuring techniques exists--a compilation of metrical data published by different authors over a long period of time fits into a general picture of morphological development of a single species, as Boessneck et al. (1971, pp. 56 ff.) have shown for domestic cattle. The general pattern of size development had been recognized earlier by scholars working independently, based mainly on their own measurements (e.g., Nobis 1954; Møhl 1957; Boessneck 1958; Jewell 1962; Degerbøl 1963; Clason 1967; Matolcsi 1970). This is only one of many examples in which measurements taken by different scholars are shown to be comparable when they are based on a common definition. The recently published measurement guide by von den Driesch (1976) with short definitions in English will help to further standardization of metrical techniques. Technical difficulties certainly do not constitute a serious argument against the application of metrical techniques in zooarchaeology.

In fact the fundamental objections of those not applying metrical analysis are not technical ones. We must demonstrate that metrical analysis is not only a descriptive method of natural history, but an analytical method of cultural history as well. The development of archaeological science has shown that the boundary between natural and cultural history is very fluid. All facts of natural history reflecting conditions which influence man or are influenced by him are useful for the reconstruction of cultural history. Bone measurements represent animal size; the size of an animal is partly dependent on its genetic pattern, and partly reflects environmental conditions. Environmental conditions are interesting in the context of cultural history. The first question therefore, seems to be--what component of animal size is fixed genetically and what component reflects environment? A general answer to this question could be given readily by animal breeders, but for zooarchaeology the issue is quite unimportant. Metrical analysis is not based on the individual animal and not even on animal populations in the sense understood by animal breeders. Even large collections of measurable bones rarely represent an average of more than one or two individuals per year of occupation of a site. The "population" being dealt with in zooarchaeology is a sample from a succession of generations. Therefore the environment will be acting not only in the ontogenesis of the individual. The influence of environment will also be reflected genetically, because the time spans involved will normally be long enough to permit complete adaptation of genetic patterns to environmental demands. In the case of domesticated animals things are more complicated; genetic and environmental conditions are not acting alone, since man exerts additional selective pressures. But in this case animal size is certainly a fact of cultural history. Some preliminary results of the author's research on sheep from the Middle East[2] illustrate how animal size can be interpreted in terms of environmental and cultural history.

The modern wild sheep of the Middle East show not only a clear difference in size between the *vignei*-sheep of the northeast and the mouflons of the west and south (held to be different species [Nadler et al. 1973]), but there are also marked size differences within the mouflons. The animals are bigger in the well-watered mountains of Turkey and western Iran, and they are smaller in hot and dry Baluchistan. This aspect clearly reflects climatic conditions, although it is impossible to say what factor(s) of climate is (are) acting. It could be temperature (Bergmann's rule), or it could be an indirect effect through a smaller alimentary base resulting from the more arid conditions. Whatever the explanation, one would expect a similar difference in size between Pleistocene and modern wild sheep. Because of the lack of extensive excavations of Pleistocene deposits within the area of distribution of wild sheep, there is a remarkable dearth of Pleistocene sheep remains. Among the collections that have been made available to me,[3] there are not enough sheep bones from such deposits to make any valid statements regarding the size of Upper Pleistocene sheep. Only from the Late Pleistocene of Palegawra Cave in northern Iraq (Turnbull and Reed 1974) comes a large enough series for study. These sheep were indeed larger than the modern forms. A certain reduction in size from the Pleistocene to the Holocene is known for most ruminant species. It is therefore less interesting to compare modern wild sheep with Pleistocene specimens than it is to compare Late Pleistocene sheep to early Holocene ones. This comparison will show whether the environmental changes were sudden or gradual. For example, Figure 1, in which two measurements of the astragalus are plotted against each other for different periods, shows no difference in size between the sheep of Palegawra and those of five other sites dated to the very beginning of the Holocene. The decline in size toward the modern wild sheep from western Iran, Turkey, and northern Iraq is clearly visible.

The conformity in size between the sheep from Palegawra and those from the upper levels of Shanidar, Zawi Chemi Shanidar, Asiab, Karim Shahir, and Mureybit is also of great importance for the question of the origins of sheep domestication. It has been argued that animal size is not a good indicator for domestication because of known size changes in wild animals (Jarman 1969). Certainly it is not correct to conclude domestication from size changes of complete populations. But, in fact, the entire population of a species was never domesticated at once. Only a part of it came under human control, and thereby changed morphologically as a result of this change in habitat. Thus it is not a simple alteration of animal size which indicates domestication, but rather the split of a population into an unaltered, presumably wild part and an altered, presumably domesticated part. As Figure 1 shows, such a split within the sheep population of the Middle East occurred sometime around 7000 B.C. Sheep domestication has already been stated for the sites being incorporated into this column of Figure 1 by Stampfli (in press) and Lawrence (in manuscript).

The issue needing discussion is whether or not there existed a phase of animal domestication prior to the morphological split. Domestication of morphologically unaltered sheep has been postulated for Zawi Chemi Shanidar (Perkins 1964). We cannot discuss this most problematical topic in detail here. Aside from the fact that the material base from which Perkins draws his conclusions is rather weak statistically, the problems of interpreting age distributions in archaeological bone assemblages have not yet been solved. Other interpretations of Perkins's finds are possible, even without having to postulate selective hunting practices to explain the high proportion of young animals. Yearlings of wild sheep are expelled by the ewes at the beginning of the lambing season (Pfeffer 1967b, p. 114). Depending on the season, and on the hunting technology, these inexperienced animals may constitute a high proportion of the game of prehistoric hunters. Thus whether or not we accept Perkins's conclusions remains largely a matter of faith.

Even if sheep at Zawi Chemi lived in some kind of closer relationship to man, this form of man-animal relationship must have been different from

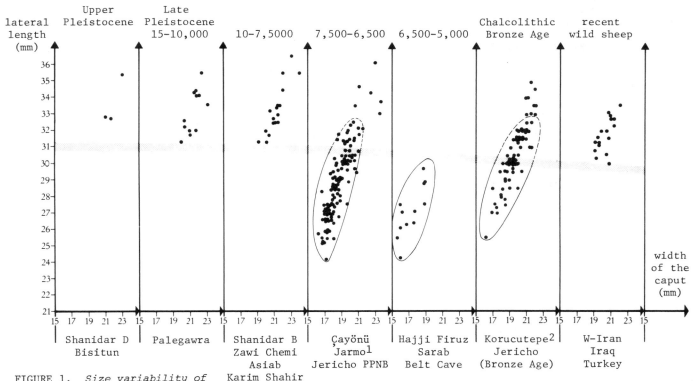

FIGURE 1. *Size variability of southwest asiatic sheep represented by measurements of the astragalus.*

1) Stampfli, in press
2) Boessneck and von den Driesch 1975

domestication of later periods. As seen through the quick-motion apparatus of archaeology the amplification of size variability at the time of early domestication occurred suddenly, and the breadth of size variation immediately reaches its full extent. As Figure 1 shows, there is no further size reduction during the seventh and sixth millennia (the lack of wild sheep remains in the period 6500 to 5000 B.C. is ascribable to insufficient faunal samples from this time; the lack of scattered points in the upper part of this column of Figure 1 cannot serve as a basis for conclusions). Throughout the rest of prehistoric times as well, we see, at present, no further increase in the variability of Middle Eastern sheep. This indicates that the full extent of alteration occurred during only one unit on the time scale of Figure 1. The man-made conditions that caused alteration in the period 7500-6500 B.C. did not cause *more* alteration in the following periods. It is therefore not logical to believe that the same conditions could have produced *less* alteration in the preceding periods. There must have been a new factor in the relationship between man and animals at the time represented by the sites of Çayönü, Jericho PPNB, and Jarmo--a factor that was not present at Shanidar B, Zawi Chemi, Asiab, Karim Shahir, and Mureybit. Whether or not one wants to call this change in man-animal relationship "the beginning of animal domestication" is a question of definition. In any case the measurable reaction of the animal population shows that by this change a stage was reached which in later prehistoric periods is called "domestication" without further qualification.

The reasons for size reduction of artiodactyls under the conditions of early domestication are still not fully understood (for a discussion and further literature see Jarman and Wilkinson 1972; see also Koch 1969), but because we have no satisfactory explanation, we should not be led to neglect the fact itself. Size reduction gives evidence for domestication in cattle and pig (e.g., Reed 1961). In the case of sheep as well, it is quite clear that human influence caused the alterations shown in Figure 1. Fundamental to this conclusion is the fact that the development of wild sheep can be followed separately, although there is an overlap with domestic sheep. For the goats of the Middle East the results of metrical analysis still have to be evaluated. Divergence of a metrically "wild" and a metrically "domestic" population seems to start at about the same time as in sheep. Whereas sheep seem to have been domesticated in the center of the Fertile Crescent, goats were domesticated in its two extremities. The situation in central Turkey is not clear. It is improbable for reasons of chronology, site ecology, and numbers of finds represented, that the sheep of Suberde really were wild as Perkins and Daly (1968) believe. The lack of published measurements prevents us from making a solid judgment regarding this site. In the southeastern part of the Fertile Crescent and its flanks towards the Iranian plateau domestic sheep at first seem to be absent. Later they are present in the upper level of Sarab, at Hajji Firuz (Meadow 1975) and throughout the Neolithic of Belt Cave. In the southwestern extremity of the Fertile Crescent remains of domestic sheep are present in the Prepottery Neolithic B levels of Jericho (Clutton-Brock and Uerpmann 1974) but their frequency is quite low.

We hope that these preliminary results suggest the potential of metrical analysis of faunal remains from the Middle East. What has been done for sheep and goats will have to be done for the other animals as well, and for all species in smaller units of time and space. But further progress in this direction will be possible only if an agreement can be made regarding more detailed publication of bone measurements. Especially among Anglo-American zooarchaeologists there is an obvious hesitation to publish measurements in detail. Bone measurements are documents and are important, whether immediate conclusions are based on them or not. Many publications on faunal remains from the Middle East do not contain any record of metrical data, and even those giving mean and variation of certain measurements are at present of little value because of the lack of comparable material published in detail. In ruminants, which have a remarkable sexual dimorphism in size, the mean values of bone measurements depend to an important extent on the proportion of male and female bones in the collection. Thus a mean value can only be representative for a population if the proportion of males and females is known. This proportion usually cannot be established. For the time being, while metrical information on the prehistoric animals of the Middle East is still virtually at point zero, it is necessary to publish in detail, bone by bone, at least the measurements of well-preserved specimens,[4] until we have a broad sample of measurements from all periods represented in the Middle East.

But we need to think further ahead already. Hundreds of thousands of bones are being dug up every year on Middle Eastern sites. Except for very important specimens, publication bone by bone will not be possible. Yet metrical information derives ultimately from the individual specimen. All abridged representations of measurements destroy the connections between the various single measurements of one bone. These connections are very important for some techniques of metrical analysis. Therefore we must find a system apart from printed publication which gives access to the information from every find. This system of course will be based on electronic data processing. Metrical data, like most data from prehistoric bone finds, are most easily processed by computers.[5] Various systems for computer processing of archaeological bone data have been developed and are in use. Based on these systems we only need to work out the problems of funding and of organization before we can create a data center for zooarchaeology, where files could be stored holding all the individual information of bone finds as postulated above. Such a file would be complementary to a printed publication. Every user of the publication would be able to ask for the raw data to verify conclusions or for comparative purposes. Many of these comparisons could be done with the help of the computer. Another result of a common data center would be progress toward further standardization of zooarchaeological methods, which is necessary for future work. We hope the development of such a center can begin before the amount of data becomes unwieldy.

Notes

1. Dr. Peter S. Wells kindly improved the English of this paper. The figure was drawn by Wolfgang Tambour.
2. This research is part of a special research project of the University of Tübingen (Tübinger Atlas des Vorderen Orients - SFB 19) which is supported by the Deutsche Forschungsgemeinschaft.
3. I wish to thank all those who have contributed to my work by giving me access to collections or by providing me with unpublished data. My special thanks go to J. Boessneck, J. Clutton-Brock, A. von den Driesch, P. Ducos, R.H. Dyson, T. Haltenorth, B. Hesse, B. Lawrence, R.H. Meadow, F. Poplin, C.A. Reed, R.L. and R.S. Solecki, H.R. Stampfli, L. de la Torre, and P.F. and W.D. Turnbull.
4. Compare with the contribution of J. Boessneck and A. von den Driesch to this volume.
5. A system ("KNOCOD") being developed and used by the author, which has been designed to include metrical information is described in more detail below.

References

Boessneck, J.
   1958   *Zur Entwicklung vor- und frühgeschichtlicher Haus- und Wildtiere Bayerns im Rahmen der gleichzeitigen Tierwelt Mitteleuropas*. Studien an vor- und frühgeschichtlichen Tierresten Bayerns, vol. 2. München.

Boessneck, J. and A. von den Driesch
   1975   "Tierknochenfunde von Korucutepe bei Elâzig in Ostanatolien (Fundmaterial der Grabungen 1968 und 1969)," in M.N. van Loon, editor, *Korucutepe* I, pp. 1-220. Amsterdam/New York.

Boessneck, J., A. von den Driesch, U. Meyer Lemppenau, and E. Wechsler-von Ohlen
   1971   *Die Tierknochenfunde aus dem Oppidum von Manching*. Die Ausgrabungen in Manching, vol. 6. Wiesbaden.

Clason, A.T.
   1967   *Animal and Man in Holland's Past*. Palaeohistoria, vol. 13A.

Clutton-Brock, J. and H.-P. Uerpmann
   1974   "The sheep of early Jericho," *Journal of Archaeological Science*, vol. 1, pp. 261-274.

Degerbøl, M.
   1963   "Prehistoric cattle in Denmark and adjacent areas," *Occasional Paper No. 18 of the Royal Anthropological Institute*, pp. 68-79.

Driesch, A. von den
   1976   *A Guide to the Measurement of Animal Bones from Archaeological Sites*. Peabody Museum Bulletin 1. Cambridge. (Harvard University).

Duerst, J.U.
1926 "Vergleichende Untersuchungsmethoden am Skelett bei Säugern," in *Handbuch der biologischen Arbeitsmethoden*, Abt. 7. Methoden der vergleichenden morphologischen Forschung, Heft 2, pp. 125-530. Berlin and Wien.

Jarman, M.R.
1969 "The prehistory of Upper Pleistocene and Recent cattle. Part I: east Mediterranean, with reference to north-west Europe," *Proceedings of the Prehistoric Society*, vol. 35, pp. 236-266.

Jarman, M.R. and P.F. Wilkinson
1972 "Criteria of animal domestication," in E.S. Higgs, editor, *Papers in Economic Prehistory*, pp. 83-96. Cambridge (at the University Press).

Jewell, P.J.
1962 "Changes in size and type of cattle from prehistoric to medieval times in Britain," *Zeitschrift für Tierzüchtung und Züchtungsbiologie*, vol. 77, pp. 159-167.

Koch, W.
1969 "Die Ernährung der Haustiere in der frühen Domestikation," in J. Boessneck, editor, *Archäologie und Biologie*. Forschungsberichte 15 der Deutschen Forschungsgemeinschaft, pp. 94-99. Wiesbaden.

Lawrence, B.
ms. "Evidences of Animal Domestication at Çayönü," in manuscript.

Matolcsi, J.
1970 "Historische Erforschung der Körpergrösse des Rindes auf Grund von ungarischem Knochenmaterial," *Zeitschrift für Tierzüchtung und Züchtungsbiologie*, vol. 87, pp. 89-137.

Meadow, R.H.
1975 "Mammal remains from Hajji Firuz: a study in methodology," in A.T. Clason, editor, *Archaeozoological Studies*, pp. 265-283. Amsterdam.

Møhl, U.
1957 "Zoologisk Gennemgang af Knoglematerialet fra Jernaldersbopladserne Dalshøj og Sorte Muld, Bornholm," *Bornholm i Folkevandringstiden*, pp. 279-318. Kopenhagen.

Nadler, C.F., K.V. Korobitsina, R.S. Hoffman, and N.N. Vorontsov
1973 "Cytogenetic differentiation, geographic distribution, and domestication in Palaearctic sheep (*Ovis*)," *Zeitschrift für Säugetierkunde*, vol. 38, pp. 109-125.

Nobis, G.
1954 "Zur Kenntnis der ur- und frühgeschichtlichen Rinder Nord- und Mitteldeutschlands," *Zeitschrift für Tierzüchtung und Züchtungsbiologie*, vol. 63, pp. 155-194.

Perkins, D.
1964 "Prehistoric fauna from Shanidar, Iraq," *Science*, vol. 144, pp. 1565-1566.

Perkins, D. and P. Daly
1968 "A hunters' village in Neolithic Turkey," *Scientific American*, vol. 219, pp. 97-106.

Pfeffer, P.
1967 *Le mouflon de Corse (Ovis ammon musiman Schreber, 1782)*. Mammalia, vol. 33, supplement.

Reed, C.A.
1961 "Osteological evidences for prehistoric domestication in Southwestern Asia," *Zeitschrift für Tierzüchtung und Züchtungsbiologie*, vol. 76, pp. 31-38.

Stampfli, H.R.
in press *The fauna of the prehistoric archaeological sites of Jarmo, Matarrah, Karim Shahir, and the Amouq in southwestern Asia*. Studies in Ancient Oriental Civilization. Chicago.

Turnbull, P.F. and C.A. Reed
1974 *The fauna from the terminal Pleistocene of Palegawra Cave*. Fieldiana: Anthropology, vol. 63, pp. 81-146.

PART THREE

DOMESTICATION AND ENVIRONMENT

The issue needing discussion is whether or not there existed a phase of animal domestication prior to the morphological split... There must have been a new factor in the relationship between man and animals at the time [7500-6500 B.C.] - a factor that was not present [earlier]. Whether or not one wants to call this change in man-animal relationship "the beginning of animal domestication" is a question of definition. In any case the measurable reaction of the animal population shows that by this change a stage was reached which in later prehistoric periods is called "domestication" without further qualification [Uerpmann, paper in Part Two of this volume].

The particular connotation of the term "domestication must vary species by species to the extent that it implies a form of socio-behavioral contract between man and individual members of a gregarious species. Domestication can be said to have taken place only when the process of animal keeping has, to some extent, resulted in changing morphological and behavioral characteristics of the animals and also in altering the socio-economic structure of the human community. Selection of particular characteristics will be determined by the nature of the interaction between species [Tosi and Compagnoni, paper in Part Four of this volume].

One approach to interpretation in zooarchaeology is to concentrate on the animals themselves - what changes in their morphology and distribution in the archaeological record suggest about the different environmental or cultural conditions under which they lived or were kept. Interaction of animal with man is viewed as a major factor in the manifestation of particular patterns, but it is the results of such interaction on the animals and on the environment which are of primary concern to the paleontologists and zoologists doing the analyses and not the inherent nature of the interaction itself. In the sense that it documents the effects of human actions, the approach can be termed culture historical. Inferences based on analogy to the morphology, behavior, distribution, and habitat requirements of modern species play an important part in this kind of analysis.

Another approach to interpretation in zooarchaeology focuses on the dynamics of the interaction between man and animal. Emphasis is put on the place of animals in different aspects of human society and on the role that human interaction with the animal world played in the processes by which such societies developed. In attempting to answer questions about the relative importance and scheduling of subsistence activities, about the role of animals in agriculture, transport, and trade, and the relationships between resource distribution and social organization, a number of prehistorians and archaeologists have turned to faunal analysis. Whether such issues can be realistically approached using animal bone data may be questioned (as, for example, by Clutton-Brock in this section) but, even so, the posing of these sorts of queries has prompted some analysts to examine faunal remains from a different perspective. All such studies must, of course, be based on a firm morphological and zoological foundation, but the search for patterns to aid in interpretation is carried far into the domain of cultural anthropology.

The two quotations from other parts of the volume were chosen to head this section because they neatly represent the two approaches to interpretation in zooarchaeology just discussed. Although neither is a particularly extreme example of one or the other point of view, they do serve to characterize different areas on a continuum. Uerpmann in his paper in Part Two looks at domestication from the point of view of the animal and its "measurable reaction" to a process which he believes took place over a very short span of time in any given locality. He rejects non-metrical and non-morphological criteria for recognizing the existence of domestication and emphasizes that, at least in the bovids, size reduction and morphological change were an integral part of domestication and could have occurred over a period so short (30 generations - ca. 60 to 150 years) that different stages cannot be separated out from an archaeological record whose smallest definable temporal units are usually more than 100 years.

For secondary domesticates such as the camel and horse, however, morphological changes were, as Bökönyi points out (this part), somewhat longer in making an appearance. Thus Ducos (this part) contends that what is important to study is changing man-animal relationships which he feels often can be documented in deeply stratified sites. In searching for methods useful for all species, he turns to a definition of domestication which emphasizes the relationship of man to individual animals and the selection process which such recognition implies. Tosi and Compagnoni (Part Four) also take this approach but emphasize that the nature of such a "contract between man and individual members of a gregarious species" must vary species by species, a point addressed by Clutton-Brock as well in her discussion (this section) of the importance of understanding the behavioral characteristics of the various ungulates present in the Middle East. Tosi and Compagnoni, however, go further than Ducos by emphasizing the importance of recognizing the changes wrought in the very fabric of society by the keeping and using of domestic animals. Again, particularly in the case of the camel and the horse, changes in society as well as in the morphology of the animal took long periods of time to become fully elaborated.

One feature which underlies both approaches, however, is their dependence on an understanding of the present to interpret the past. Indeed, such inference by analogy, whether explicit or (more commonly) implicit, is found in all aspects of zooarchaeological research. The logic employed is basically the following - if two things, one from the present and one from the past, agree in one or more respects, they will probably agree in other respects. The degree of probability will vary with the level of abstraction involved in making the analogy. Since zooarchaeologists deal in part with biological phenomena, there is a strong component of uniformitarian justification employed in such reasoning, i.e., that the same physio-chemical and biological processes which operate today operated also in the past. Thus the bones of modern animals of known species, age, and sex are used as a basis for identification of ancient remains. Studies of the physiology, behavior, habitats, and distribution of wild and domesticated species are examined with the aim of identifying ranges of likely variation for each form; the tendencies and limits so established are employed to interpret past configurations. Reports on man-animal relationships in traditional societies are scrutinized with an eye toward determining underlying patterns in terms of which the archaeological record can be elucidated. Particularly at the interpretative level, however, the use of the present to look at the past must be done with great care, with all due regard for changes which may have taken place in the meantime and most specifically with respect for the changes wrought by human societies. Nowhere is this more true than in the analysis of past environments.

The presence of certain taxa in archaeological deposits can be used as a guide to construction of a picture of past environmental conditions only after a number of complicating factors have been sorted out. Bökönyi's contribution to this section demonstrates how the remains of larger animals in a site can sometimes be employed as general indicators of the nature of the landscape exploited by man and, indeed, the very nature of that environment will help determine the kind of exploitation possible. It is the use of these animals by man, however, that is reflected in the archaeological record and thus the disappearance of larger animals from that record need not reflect any change in the nature of the physical environment but only in man's interaction with it. Small mammals, amphibians, reptiles, and birds often have more sharply circumscribed habitat requirements than do larger animals but these are understood in detail for only a few species. In circumstances where the remains of such taxa end up in a site through agencies independent of direct human action, they can be used in the fashion carefully outlined by Redding (this section) to monitor environmental changes in the area immediately around the site.

In concluding the introduction to this section, we turn to the paper by Zeder. Alone of all the contributions in this volume, it deals not with archaeological but rather with modern faunal material. To be sure, the original stimulus for the study was an attempt to duplicate on specimens from animals of known relationships with man and the environment results originally observed on archaeological remains. The study grew, however, into a first step in trying to deduce where and how to look for morphological distinctions within the bones of herd animals given differences in the interaction of these animals with both the physical environment and man. Application of the results of such studies to the interpretation of the archaeological record will ultimately depend on the ability to progress beyond characterization to explanation. For modern phenomena are only a starting point against which to assess past patterns; correspondences found and analogies drawn must not be viewed in themselves as explanations for these patterns.

R.H.M.
M.A.Z.

BONES FOR THE ZOOLOGIST

Juliet Clutton-Brock, British Museum (Natural History)
London

Archaeologists are increasingly accused of failing to make use of the rich biological potential that is brought to light by their excavations. Being of course primarily interested in the history of human cultures they retaliate by asking the biologists to supply them with ever more detailed information on the agriculture, diet, and way of life of the human populations whose remains they are finding. The archaeologist wants to know what the people ate, whether the animals were hunted, herded, or domesticated. How many animals were killed at what time of year, how much meat was provided by each carcass, and even, what was the probable daily intake of meat of each inhabitant of the site. As the questions become more elaborate so the techniques used for their elucidation become increasingly refined. The result is an edifice of speculation and guesswork that is usually accepted without doubt by the archaeologist. But is any zoologist well enough trained in economics, geography, mathematics, demography, and livestock husbandry to pronounce with authority on all these quandaries? The latest technique to be developed is that of site-catchment analysis which is based on a territorial approach to archaeology (Vita-Finzi and Higgs 1970; Higgs and Jarman 1975). Geographical and geological surveys are made of the area and the ecological and resource potential is assessed from investigations into the probable landscape, climate, fauna, and flora of the site at the time of occupation. Ethnographic reconstructions are employed to calculate the probable numbers of people inhabiting the site and their subsistence level, so that, when carried to its most esoteric extremes, it becomes no longer necessary to excavate the site. I actually believe that this method is as likely to produce results no further from the truth than traditional methods of excavation, with its often random collection of animal remains, calculation of minimum numbers of individuals, meat weights, relative proportions of the species of livestock present and the rest of the components expected of the usual "animal bone report."

It may be questioned that if the zoologist is not qualified to give the archaeologist the answers to his questions, who is? I believe that some of these problems are really insoluble by a zoologist from a study of nothing but bones and teeth. In an attempt to prove that environmental studies should be a principal facet of archaeology the zoologist has been pressed into trying to squeeze too much information out of the available sources of evidence. There is a danger that as a result the subject will lose integrity. If the zoologists were the excavators and they parceled off the sherds and artifacts to the archaeologist, simply for evidence of dating, a faunal history of the site might emerge that would be very different from traditional human cultural history with all its ramifying channels of enquiry. It is the process of these faunal changes and the zoological results from man's shifting role as a predator who was just one member of a balanced ecosystem to a unique mammal in control of his environment that I find most intriguing. Of course, the basis of this work must lie in the correct identification of the species of animal whose remains are found on archaeological sites, identification to the genus level is not sufficient, and in accurate assessments of the relative abundance of the species represented. But I prefer to use these identifications in the discussion of their zoological implications rather than in the perhaps spurious reconstruction of past human economic systems. There is, however, another more theoretical topic that I am not immune to speculation on and which I should like to discuss briefly here. This concerns the reasons why so few species of mammal have been domesticated and why, if the archaeological record is correct, domestication first occurred in western Asia.

The large collections of mammalian remains from early prehistoric sites in western Asia have provided, and I hope will continue to provide, prolific evidence for the early history of domesticated animals and the changes that occurred in faunal assemblages as man became the dominant species. It has to be remembered, however, that from every point of view western Asia is a most complicated region with four well-defined environmental zones: the coastal wooded Mediterranean strip, the Jordan rift valley, the arid steppe zone, and the high mountain and plateau areas. Moreover, besides having these varied environments the region, throughout the Quaternary, has been a junction for biological invasions from tropical

Africa, the Mediterranean, the Middle Eastern alpine zone and from eastern Asia. Any examination of the changes in faunal history that have occurred since the end of the Pleistocene have therefore to be considered in the context of the mixing of climatic and biotic zones. In the early Holocene the archaeological evidence suggests that a slowly increasing human population moved about between the environmental zones and learned how to colonize the less hospitable areas by manipulating whatever natural resources were available.

A characteristic of the hunting peoples of the early Holocene was that they had a broad-based economy that relied on a large number of varied food resources (Flannery 1969, p. 77). As the Natufian gave way to the Neolithic a higher proportion of meat was obtained from single species of wild game. Within each zone of western Asia the species differed, so that in southern Anatolia on such sites as Çayönü (Braidwood et al. 1974) and Çatal Hüyük (Perkins 1969) the majority of the animals killed were wild cattle, red deer, sheep, and goat. In the Levant the greatest number of animal remains came from gazelle with fox also an important source of meat (Clutton-Brock 1969, 1971). At Jarmo (Reed 1959, 1960) and in the Zagros Mountains (Hole et al. 1969) wild sheep and goat predominated. It was at this period that the fundamental change in the human way of life from hunter to farmer became established. Quite why it occurred is an endless source of argument and discussion which I will not reiterate here except to say that the theory first propounded by Harlan and Zohary (1969) that the earliest cultivation of cereals occurred in areas somewhat away from where the wild grain grew in natural abundance, with the need for food storage being the initial stimulus, seems eminently reasonable. In this way new land could be colonized with, in the next stage, captive livestock providing an insurance against a lack of meat in lean times.

The much quoted work of Lee (1968) on the hunting practices of the !Kung bushmen has shown that, at least in areas of low human population density, hunter-gatherer communities can obtain sufficient food for adequate subsistence with remarkably little effort. It follows that some environmental stress must occur to make people change to the life of drudgery that follows from the sowing of seeds; for growing crops have to be watered, manured, weeded, and protected from marauding ungulates. It is now generally believed that agriculture preceded the domestication of livestock and it seems to me that Zeuner's theory of the "crop-robbers" (1963, p. 199) may indeed have much relevance to the history of its earliest phases. In discussion with Pierre Ducos, we agreed that the building of walls and fences by primitive peoples is much more likely to be done in an attempt to keep animals away from the crops rather than to contain livestock. There are many accounts of the lack of fear shown by wild ungulates towards man when they have not been hunted with modern weapons, and I think that there would have been a much closer physical relationship between man and the wild ungulate populations of western Asia in the early Neolithic than is generally assumed. It would have been a very easy task for early man to tame any of the species of large herbivores in his surroundings but I believe he may have been more interested, in the early stages of agriculture, in driving them away from his settlements.

Contrary to the beliefs of Zeuner (1963, p. 434), Legge (1972, pp. 119-124), and others, I do not believe that man exercised any direct control over the animals that he hunted; I think it improbable that he herded them, kept them in captivity, or culled them in any programmed way. Gazelle were extremely common in western Asia at the end of the Pleistocene, as red deer were in Europe, and it is clear that these species were a primary source of meat, perhaps for thousands of years, but there is no osteological or ethnographic evidence to suggest that man had any direct control over them. Although no doubt the hunters were aware of the territory occupied by the wild herds and made use of known trails and drinking places where the animals were commonly to be found. Probably some awareness of the need for restraint and management in the way the animals were exploited was also part of the hunters' way of life, as described by Jewell (1974).

The clues to the history of domestication must be looked for in the behavior patterns of the potential domesticates as well as in those of the hunters. It is often said that it is surprising that so few of the large ungulates have been domesticated and incidentally it is certainly true that a new look will have to be taken at these wild species as an untapped food resource (Jewell 1969). But in the prehistoric period only certain groups could flourish under the drastic alterations to their way of life that dominance by man entailed. Experiments with eland, oryx, red deer, and African buffalo have shown that these animals can be tamed and successfully herded at the present day by shepherds who stay with the free-ranging herds, but I think it unlikely that these animals would have bred successfully under the harsh conditions of primitive domestication (see Jewell 1969 for references, also Blaxter et al. 1974). Gazelle and many species of antelope are strongly territorial in their behavior and when feeding they range over a wide area and have a high inter-individual distance, that is they cannot be herded close to each other. Domestic livestock have to be penned at night, if not also tethered by day, and this would result in loss of night feeding which is crucial to the well-being of many species. Sheep, and particularly goats, are flexible in their feeding habits. They are not markedly territorial, their social structure being based on well-established dominance hierarchies in the male and less pronounced hierarchies in the female, and they have a short inter-individual distance so they can be bunched together (Geist 1971; Jewell et al. 1974). All these characters make sheep and goats better suited to being dominated by man than gazelle would be. Even so the dramatic decrease in size that can be seen between the bones of wild, hunted animals and their small domesticated counterparts must be the direct result of constraint and malnutrition imposed on the captive animals. There is little experimental information on the effects of varying nutrition on the breeding capacities of ungulates (Sadleir 1969) but it is probable that periods of malnutrition would reduce fertility. This could be expected to be especially marked in the males of territorial species such as deer and gazelle and could be further reduced by stress and by the vitamin and mineral deficiencies that would be the inevitable result of restricted feeding.

By 7000 B.C. the village farmers of western Asia had become dependent on cultivated cereals and on the species of domesticated livestock that still support the human communities at the present day. The reasons why only this small group of ungulates was domesticated and no others may be surmised, but it must now be time for expansion. The untapped resources of meat that could be made available by modern methods of management and game cropping of wild African ungulates are only just beginning to be appreciated. A proper understanding of the history of domestication and of the behavior of the dwindling stock of wild herbivores will be a considerable aid to man's further manipulation of his environment.

References

Blaxter, K.L., R.N.B. Kay, G.A.M. Sharman, J.M.M. Cunningham, and W.J. Hamilton
1974   *Farming the Red Deer*. Edinburgh.

Braidwood, R.J., H. Cambel, B. Lawrence, C.L. Redman, and R.B. Steward
1974   "Beginnings of village farming communities in southeastern Turkey - 1972," *Proceedings of the National Academy of Sciences,* vol. 71, no. 2, pp. 568-572.

Clutton-Brock, J.
1969   "Carnivore remains from the excavations of the Jericho tell," in P.J. Ucko and G.W. Dimbleby, editors, *The Domestication and Exploitation of Plants and Animals,* pp. 337-345. London.
1971   "The primary food animals of the Jericho Tell from the Proto-Neolithic to the Byzantine period," *Levant,* vol. 3, pp. 41-55.

Flannery, K.V.
1969   "Origins and ecological effects of early domestication in Iran and the Near East," in P.J. Ucko and G.W. Dimbleby, editors, *The Domestication and Exploitation of Plants and Animals,* pp. 73-100. London.

Geist, V.
1971   *Mountain Sheep: A Study in Behavior and Evolution*. Chicago.

Harlan, J.R. and D. Zohary
1966   "Distribution of wild wheats and barley," *Science,* vol. 153, no. 3740, pp. 1074-1080.

Higgs, E.S. and M.R. Jarman
1975   "Palaeoeconomy," in E.S. Higgs, editor, *Palaeoeconomy,* pp. 1-7. Cambridge (at the University Press).

Hole, F., K.V. Flannery, and J.A. Neely
1969   *Prehistory and human ecology of the Deh Luran Plain*. Memoirs of the Museum of Anthropology, University of Michigan, no. 1. Ann Arbor.

Jewell, P.A.
1969   "Wild animals and their potential for new domestication," in P.J. Ucko and G.W. Dimbleby, editors, *The Domestication and Exploitation of Plants and Animals,* pp. 101-109. London.
1974   "Managing animal populations," in A. Warren and F.B. Goldsmith, editors, *Conservation in Practice*. London.

Jewell, P.A., C. Milner, and J. Morton Boyd
1974   *Island Survivors: The Ecology of the Soay Sheep of St. Kilda*. London.

Lee, R.B.
1968   "What hunters do for a living, or, How to make out on scarce resources," in R.B. Lee and I. DeVore, editors, *Man the Hunter,* pp. 30-48. Chicago.

Legge, A.J.
1972   "Prehistoric exploitation of the gazelle in Palestine," in E.S. Higgs, editor, *Papers in Economic Prehistory,* pp. 119-124. Cambridge (at the University Press).

Perkins, D.
1969   "Fauna of Catal Huyuk: evidence for early cattle domestication in Anatolia," *Science,* vol. 164, no. 3876, pp. 177-179.

Reed, C.A.
1959   "Animal Domestication in the prehistoric Near East," *Science,* vol. 130, no. 3389, pp. 1629-1639.
1960   "A review of the archaeological evidence on animal domestication in the prehistoric Near East," in R.J. Braidwood and B. Howe, editors, *Prehistoric Investigations in Iraqi Kurdistan*. Studies in Ancient Oriental Civilization no. 31, pp. 119-145. Chicago.

Sadleir, R.M.F.S.
1969   *The Ecology of Reproduction in Wild and Domestic Mammals*. London.

Vita-Finzi, C. and E.S. Higgs
1970   "Prehistoric economy in the Mount Carmel area of Palestine: site catchment analysis," *Proceedings of the Prehistoric Society,* vol. 36, pp. 1-37.

Zeuner, F.E.
1963   *A History of Domesticated Animals*. London.

"DOMESTICATION" DEFINED AND METHODOLOGICAL APPROACHES TO ITS RECOGNITION IN FAUNAL ASSEMBLAGES

Pierre Ducos, Centre national de la Recherche scientifique (E.R. 166)
Centre de Recherche d'Ecologie humaine et de Préhistoire
Saint-André-de-Cruzières

In approaching the analysis of faunal remains from early settlements in the Middle East, the central problem is "what data can we use to decide if one animal species is domestic or wild?" This question should be among the first asked because, in this area where we find the oldest village communities (as well as the first evidence of the manufacture of pottery and the use of metals), we expect also to find some of the earliest remains of domesticated animals. In order to obtain a clear idea of the history of domestication in the area, we must have a means by which to decide whether or not what we generally call domestication was practiced by the inhabitants of a given settlement. But we cannot provide an answer to that question so long as we have no adequate, that is to say comprehensive, definition of domestication which is useful when analyzing archaeologically derived faunal materials. In my opinion, on the one hand, a well-grounded definition must be devoid of *a priori* causes, mechanisms, or consequences of the very process we are studying; those features of domestication are precisely what we have to demonstrate. On the other hand, the definition must be useful in that a practical method must exist for recognizing domestication as defined. Finally, the proposed definition must include all known situations that are intuitively felt to be domestication.

It is not obvious, however, that there does exist a single common criterion for all the man/animal relationships we call domestication. In fact it is possible that our intuition of what is domestication corresponds to modern situations, not to ancient ones. Domestication today could appear to be a specific relationship between man and some animal species just because other patterns of relationships have disappeared. Some think it best to take an ecological approach, examining the interactions between the species to construct a definition of domestication. The interaction between man and the species exploited are considered within the ecosystem to which both belonged along with all other species sharing the same environment. These relationships are described by reference to energy flow within the ecosystem with "man being considered as a subsystem coextensive with the exploited system" (Margalef 1968). But, using this approach, it is apparent that the man/dog relationship is very different from the man/horse relationship. It is thus doubtful that a comprehensive definition of domestication can result from such an ecological approach, with the additional problem that, should we use such a definition for historical purposes, we would be describing in an *a priori* fashion the mechanisms of domestication.

Other definitions proposed take into account not the relationship itself but its secondary effects. Wilkinson (1972) reserves the term "domestication" for situations in which the seasonal subsistence cycle of the species involved is changed by man to coincide with the requirements of the human group. This definition is not adequate for all situations which are intuitively recognized as domestication (the cat, for example). Although Wilkinson may be starting from an actual case, that of the musk-ox, the typology of man/animal relationships he proposes is really based not on an observed situation (cf.ibid., p. 33) but upon theoretical considerations. His typology is *a priori* and not deduced from observed data, the same being true of his suggestions as to the causes of domestication (exploitation).

The classic definition of domestication given by Bökönyi (1969) presents: "...the essence of domestication as the capture and taming by man of animals of a species with particular behavioral characteristics, their removal from their natural living area and breeding community, and their maintenance under controlled breeding conditions for profit." Such a definition provides, in advance, answers to the question "how did man come to use domesticated animals?" ("by capture") and to the question "what for?" ("to control the animals"). But is it possible to say that domestic animals are everywhere kept in captivity, a condition which supposes that man keeps a constant watch over them, a relaxation of which would permit the animal to move out of the scope of the human group? Such a situation of constant vigilance is not found with cattle, pigs, sheep, or goats in western Africa, for example, nor with horses and bulls in the French Camargue. It is excessive to see in Bökönyi's definition the "essence of domestication." Furthermore, it is well known that the criterion of reproduction in captivity or under the control of man cannot be applied to all cases of domestication. This "classic definition," then, is replete with *a priori* propositions on the causes, mechanisms, and consequences of domestication.

In point of fact, any definition of domestication which takes the animal as the basis for the formulation introduces such *a priori* assumptions. For all those features which distinguish the domestic animal from the wild one (whether biological or behavioral) stem not from the evolutionary dynamics of the animal but from those of the human society. Therefore, that which is characteristic of domestication must be defined with reference to human society. With domestication, it is clear that "de nouveaux rapports se sont établis entre l'homme et l'animal, d'un type 'amical', et qui ne sont pas sans rappeler ceux que les hommes entretiennent à l'intérieur d'un groupe" (Haudricourt 1962). Therefore, I suggest the following definition of domestication: domestication can be said to exist when living animals are integrated as objects into the socioeconomic organization of the human group, in the sense that, while living, those animals are objects for ownership, inheritance, exchange, trade, etc., as are the other objects (or persons) with which human groups have something to do. Living conditions are among the consequences of domestication, not the mark of it.

Such integration of the animal into the human group introduces for man the necessity of being able to distinguish, within the domestic animal population, between groups of a few individuals and usually to recognize individual animals. In hunting, man/animal relationships are between populations; with domestication these relationships tend to be between individuals or between an individual and a group of individuals. In the case of the principle food species, each individual among the domestic population is generally recognized by man who may choose whether to kill or castrate or otherwise exploit the animal. So a situation is introduced in which man is able to control the animal population, but I feel that while such control is possible under conditions of domestication, it is not necessarily one of the causes for the appearance of domestication.

Using this definition, it is possible to devise a method for recognizing domestication in the zooarchaeological record (at least in those cases where it existed in the sense of our definition). A working hypothesis, based on the particular nature of faunal collections is as follows: there is a correlation between some numerical characteristics of the set of bones of animals killed by man and the techniques of acquisition and utilization of those animals by man. The reality of the hypothesis is verified if numerical data from collections representative of the same techno-economic stages in the same area appear to be closely related. The conditions are met, for example, in successive strata of the same archaeological period at a given site. At Tell Mureibet, for instance, we have a succession of seventeen levels throughout which numerical data such as frequencies of different age groups remain more or less constant.

The question that must now be put is: what configuration of the data will permit a decision to be made concerning whether animals were acquired through hunting or husbandry? The answer to this question lies in demonstrating the possibility of choices having been made—choices which were the consequence of needing to recognize individual animals and to integrate them as objects into the socioeconomic structure of the human group. We can demonstrate the existence of choice by man between individuals within an animal population with the view of their use for food if, after dividing the animal population into categories, we can reject the null hypothesis: "there is no choice between those categories." Although it is not possible to speak here in greater detail about the practical application of the method (see Ducos 1968, pp. 1-19; 1973; 1975), it is important to note that any degree of choice which can be demonstrated to have existed is not, by itself, a criterion for domestication. Its correlation with domestication depends upon both the patterns produced by such choice and the categories of animal concerned. This can be illustrated by the following example.

At Ain Mallaha (Natufian period, Palestine), it can be shown that the distribution of age groups of hunted gazelles presents a significantly different pattern than does the age-group distribution in natural populations (Ducos 1968, p. 73). Against an absence of 0-1 year-old gazelles, it can be demonstrated that there was no choice between the age groups 1-3, 3-5, 5-7, and 7+ years. This particular distribution is unlikely to be accidental (χ^2 very significant) and we must conclude that the choice was between two categories: the young less than one-year old on the one hand and the animals more than one-year old on the other. Since animals less than one-year old can easily be separated from older animals by their size alone, it was not necessary to have knowledge of the gazelle population individual by individual to have been able to make this distinction. Thus such choice cannot be cited as evidence for domestication; in fact the opposite conclusion is indicated.

The opposite sort of numerical data has led Legge (1972) to propose that domestication of gazelle can be demonstrated for the Prepottery Neolithic period. The frequency of immature metapodials suggests to him that there was an abnormally high frequency of animals from the 0-2-year age group. My own data show that the frequency of that age group in a natural population would be 46 percent. There is a frequency at Nahel Oren of over 50 percent. Tested statistically, however, the χ^2 would need to be greater than 6.64 for the Natufian levels to be judged significantly different (p=0.01) from the wild standard. The actual χ^2 is, in fact, 5.53 suggesting that the hypothesis of gazelle domestication cannot be supported. A similar demonstration can be made in the case of the sheep at Zawi Chemi Shanidar.

"Domestication" Defined   55

Age Groups (months)	Ain Mallaha (Natufian)	Mureybit I-X	Mureybit XIII-XVII	Munhatta 5-6	Abou Gosh	Beisamun	Hagoshrim	Munhatta 2	Safadi	Metzer	Gat	Nagila	Naharya	Reference Frequencies
			Prepottery Neolithic				**Pottery Neolithic**		**Chalcolithic**		**Bronze Age**			
Sus	%					%				%				%
0–6	12.5					5.1				16.7				14.4
6–12	15.6					12.7				40.0				12.4
12–24	20.3					15.2				14.3				9.9
24–60	3.7					6.2				1.9				5.7
60+	0.9					1.5				0.9				1.9
N	32					79				60				
chi^2	10.3					10.0				57.6				
Bos		%	%		%	%	%				%			%
0–24		56.3	26.0		55.6	31.7	9.7				30.0			26.8
24–48		15.6	27.1		15.9	19.5	12.9				12.5			19.9
48–66		2.5	24.2		10.2	27.3	38.7				26.0			14.2
66–78		5.0	15.7		8.9	6.8	23.2				12.0			9.9
78–138		5.0	0.0		1.3	3.0	0.0				8.0			6.8
138+		3.1	0.6		0.8	.6	0.0				0.0			3.7
N		32	46		63	82	31				40			
chi^2		15.6	9.4		27.7	25.0	<u>40.0</u>				12.3			
Ov/Cap				%	%	%	%	%		%	%	%		%
0–12				18.2	10.7	25.9	8.3	38.7		23.9	21.2	36.4		38.3
12–24				30.3	45.2	32.8	36.5	40.1		39.4	23.3	38.7		24.5
24–36				34.6	36.9	32.8	46.9	16.1		28.4	33.3	18.5		15.6
36–48				15.2	7.1	5.2	8.3	3.6		8.3	16.7	3.4		9.9
48+				0.0	0.0	0.9	0.0	0.5		0.0	1.4	0.8		3.9
N				33	84	58	96	137		106	90	297		
chi^2				13.9	<u>66.3</u>	19.9	<u>94.1</u>	31.3		<u>33.5</u>	<u>43.0</u>	<u>58.2</u>		

TABLE 1. Age-group distributions for Near Eastern settlements

The best argument for the significance of age-group distributions is given by the set of results from the settlements of the Near East which are summarized in Table 1. The values of χ^2 are calculated using the age-group distributions observed for each species at each site and an age-group distribution (called the "reference frequency" in the table) which is an estimate of the natural distribution for each species.

The age-group distributions for cattle show that there could not have been domestication-related choices being made earlier than the Pottery Neolithic. One notable aspect of the age-group distributions for cattle from Prepottery Neolithic sites is the general abundance of 0-2-year-old individuals. This distribution can be explained by the fact that young animals are easier to hunt than are adult animals. (In all of the Neolithic sites, Bos are very large.) The distribution of cattle in the Bronze Age site of Gat is very similar to the natural one because the two sexes were probably being slaughtered at different ages, the females being kept into adulthood.

For sheep and goats, domestication can be expected to have occurred as early as the Prepottery Neolithic period (at Abou-Gosh, Ramad, Beisamun, and Munhatta). At Abou-Gosh (PPNB), 50 percent of the bones from economically important animals are those of a goat of a size similar to that of Capra aegagrus. The frequency of age classes shows that, in fact, individuals were being selected to be slaughtered. But one is able to demonstrate, taking into consideration both frequency of males with respect to females and the behavior of Capra, that such individual choice could have been carried out on wild herds. In this case, there would be a situation of proto-stockrearing of a kind which Leroi-Gourhan has defined: relationship of a husbandryman to an animal maintained in its biotope and natural state of behavior. Domestication before the Chalcolithic was only for the production of meat, and young individuals were generally not killed. From the Chalcolithic period, male and female animals were treated in different ways (as with cattle).

To summarize the results, we can say that before the Pottery Neolithic period in Palestine and Syria, there was domestication only of Ovis and Capra with cattle being added in the Pottery Neolithic. Domestication was used only for production of meat until after the Chalcolithic when cattle, sheep, and goats were probably used also for the products of the living animal (milk, hair, etc.) By this time as well, domesticated animals had become different in their morphology from their wild ancestors.

I would now like to elaborate on my definition of domestication. The word "integration" allows us to presume varying degrees of "domestication." If the integration of a living animal into the socioeconomic system of a human group is to be a characteristic of all patterns of domestication, then one can expect that the degree of such integration could be variable. There are, in my definition, no postulates about progression or about the suddenness of the appearance of domestication just as there are no postulates about the causes or the nature of the outcome of the process. All these questions must be answered from objective data. The statistical method or any other suitable mathematical method can be applied.

In a similar fashion, the biological consequences of domestication on a particular species, especially those relating to changes in morphological characteristics, should not be presupposed but must be demonstrated. We cannot positively state that "there is necessarily a long time from the beginning of domestication and the appearance of morphological differences between wild and domestic animals of the same species." We can only propose that this may have been so, with the result that animals without morphologically distinguishing characters could be present in collections from the early Neolithic periods.

Some zooarchaeologists interested in morphological characteristics of domestic animals may find my definition ambiguous because it does not refer to biological criteria. Using my definition, a domesticated animal can be similar genotypically or phenotypically to a wild animal of the same species. Therefore, I think that our terminology would be clearer if we used the word "domestic" (and not "domesticated") only for specific animal populations which present well-defined morphological characteristics not found in wild populations. Thus, the elephant, reindeer, and bee would be "domesticated" animals while cattle, dog, or ferret would be "domestic" animals.

In conclusion, I would insist upon one point. The contribution of zooarchaeology to the knowledge of the prehistoric civilizations of the Near East could be of a different order than it is, for example, in Europe. In the Levant, we can trace 5000 years of successive stages of social organization and economy of human communities. These have in common not the production of domestic plants and animals so much as that they can be defined as villages. In such a context, zooarchaeology must try to provide some answers to the problems of consumption, distribution, and acquisition of food as well as to problems associated with the environment, social organization, economy, and demography. Zooarchaeology is able to contribute answers to these questions in part by the analysis of numerical data. This is not to minimize the importance of the more properly "zoological" aspects of our research, but in the case of oriental civilizations, these represent only the first part of our work. For the second part I feel that, especially for animals of food value, it is indispensable to test their status as wild or domesticated.

References

Bökönyi, S.
   1969   "Archaeological problems and methods of recognizing animal domestication," in P.J. Ucko and G.W. Dimbleby, editors, *The Domestication and Exploitation of Plants and Animals*, pp. 219-229. London.

Ducos, P.
   1968   *L'origine des animaux domestiques en Palestine*. Publications de l'Institut de Préhistoire de Bordeaux, Mémoire 6.
   1973   "La signification de quelques paramètres statistiques utilisés en Palethnozoologie," *L'Homme, hier et aujourd'hui*, pp. 307-316. Paris (Editions Cujas).
   1975   "Analyse statistique des collections d'ossements d'animaux," in A.T. Clason, editor, *Archaeozoological Studies*, pp. 35-44. Amsterdam/New York.

Haudricourt, A.
   1962   "Domestication des animaux, culture des plantes et traitement d'autrui," *L'Homme*, vol. 2, no. 1, pp. 40-50.

Legge, A.J.
   1972   "Prehistoric exploitation of the gazelle in Palestine," in E.S. Higgs, editor, *Papers in Economic Prehistory*, pp. 119-124. Cambridge (at the University Press).

Margalef, R.
   1968   *Perspectives in Ecological Theory*. Chicago.

Wilkinson, P.F.
   1972   "Oomingmak: a model for man-animal relationships in prehistory," *Current Anthropology*, vol. 13, no. 1, pp. 23-44.

ENVIRONMENTAL AND CULTURAL DIFFERENCES AS REFLECTED IN THE ANIMAL BONE SAMPLES FROM FIVE EARLY NEOLITHIC SITES IN SOUTHWEST ASIA

Sándor Bökönyi, Archaeological Institute of the
Hungarian Academy of Sciences
Budapest

From Southwest Asia, the area of earliest animal domestication (according to our present understanding), identified and published animal bone assemblages are not at all plentiful. Particularly rare are samples which come from well-dated sites, which were collected carefully and level by level, and which are sufficiently large to permit the zoologist to draw reliable conclusions supported by significant statistical evidence. The small number of such samples is even more striking if one considers both the size of the region in question and the number of different environmental types and cultures represented within it. To reconstruct any satisfactory kind of faunal history of the region is impossible without having a considerable number of animal bone assemblages from all of the environmental types. The only way to accumulate such material is by zoologists cooperating closely with excavating archaeologists and identifying the bone samples collected by them. This paper presents the results of preliminary studies on the animal bone samples from five early Neolithic sites of Southwest Asia and attempts to demonstrate how cultural and environmental factors left different marks on each assemblage. These bone samples were collected with special care and are large; in fact all are above the safety limit of five hundred identified specimens (not including whole skeletons or articulated units). Therefore, the information given by the assemblages should be reliable.

All five sites whose bone samples are discussed here were on the level of the "primary effective village farming community" (following Braidwood's [1969] terminology) that corresponds approximately to the end of the Prepottery Neolithic or to the beginning of the Pottery Neolithic of the region. By this stage, settlements obtained the major portion of their food requirements through agricultural production, although hunting and gathering remained important in some localities. From a zoological point of view, the main characteristic of this stage is that although man already possessed either all five Neolithic domestic species—cattle, sheep, goat, pig, and dog—or at least four of them (cattle or pig could be missing depending upon the geographical situation of the sites), nevertheless the importance of animal husbandry did not decisively exceed that of hunting in all regions.

The first site to be considered is that of Labweh in Lebanon which was excavated by Diana Kirkbride. It lies on the fertile plain of the Beka'a Valley between the Lebanon and the Anti-Lebanon Mountains not far from the famous classical site of Baalbek. The archaeological assemblage represents the earliest Neolithic, and its radiocarbon dates extend from 6000 to 5750 B.C. (all uncorrected dates). The excavations yielded 1043 identified animal bones.

The second site, Umm Dabaghiyah, was also excavated by Kirkbride. The site is comparatively small (ca. 100 m by 85 m by 4 m high) and was almost completely excavated. It lies in the Jezira, a somewhat hilly but treeless salty steppe between the Tigris and Euphrates rivers south of Mosul and the Jebel Sinjar (Mountains) and about thirty kilometers west of Hatra. In the region there are wells, ponds, and watercourses which are all saline, a fact which does not make the area particularly suitable for permanent settlement today. The recent inhabitants of the region are nomadic Bedouin who practice a rather primitive form of animal husbandry. The uppermost level of Umm Dabaghiyah is assigned to Hassuna Ia, a period which represents the earliest Pottery Neolithic of central Iraq, while the lower levels go back into the Prepottery Neolithic. Radiocarbon dates range between 5800 and 5300 B.C. (Kirkbride 1972). Out of the extremely rich animal bone assemblage, some 19,000 specimens could be identified.

The third site, Tell es-Sawwan, lies immediately on the left bank of the Tigris River just south of Samarra. Its archaeological material dates from the Hassuna and Samarra periods, i.e., from the end of the sixth and beginning of the fifth millennium B.C. Excavations had been going on for several seasons when, in 1972, the energetic young

58 *Approaches to Faunal Analysis*

Iraqi archaeologist, Walid Yasin al-Tikriti, took over direction and started to collect carefully the animal remains, recovering as many as 3073 identifiable specimens in a single season (to be compared with the ca. 200 specimens collected during the previous seven seasons).

The fourth site is Choga Mami, located at the Iraqi-Iranian frontier in the foothills of the Zagros Mountains at about the same latitude as Baghdad. The site was excavated by Joan Oates. Of the rich bone sample, only those from the earliest levels belonging to the Samarra period will be considered in this paper--some 646 identified specimens.

The fifth site is Tepe Sarab located in the Kermanshah Valley of western Iran and excavated by Robert J. Braidwood. The site lies in a rather narrow valley high in the mountains and has yielded an assemblage of artifacts which are reminiscent of that from Jarmo, although perhaps typologically more advanced (Braidwood 1960). Sarab can be dated to the sixth millennium B.C. The excavations yielded 8382 identified animal bone remains (Bökönyi 1977).

As the above descriptions of the sites indicate, they lie in five distinctly different types of environment: 1) fertile plain between two mountain ranges; 2) dry, saline steppe; 3) riverine oasis; 4) dry, hilly country; and 5) narrow valley high in the mountains.

With regard to their vertebrate fauna, the five sites can be divided into two groups. To the first group, Umm Dabaghiyah alone belongs; the other four sites together form the second group. The basis for this division is the ratio of domestic animals to wild animals identified in the faunal sample (Table 1 and Fig. 1)--in other words, the relative importance of animal keeping as compared to hunting for filling the animal protein needs of the ancient inhabitants. At Umm Dabaghiyah, the domestic ratio is very low (Bökönyi 1973c) being just over 11 percent, while at the four other sites remains from domestic animals dominate in an overwhelming fashion those from wild animals, being 90 percent at Tell es-Sawwan, 89 percent at Choga Mami, 84 percent at Labweh, and ca. 64 percent at Sarab.

The animal bone sample from Umm Dabaghiyah is of a unique type. Neither in Southwest Asia nor in Southeast Europe has there yet been found any other site with a fauna comprised of all five domestic species of the Neolithic yet with such a low domestic ratio and with the onager remains being so frequent (66 to 70 percent) in the wild fraction. The high domestic ratio of the second group of sites is reminiscent of that from the earliest Neolithic sites in Greece (Boessneck 1962; Higgs 1962; Jarman and Jarman 1968; Gejvall 1969; Bökönyi 1973b) and in south Yugoslavia (Bökönyi 1976). Note however that the European sites predate those of the Middle East by 500 to 1500 years.

FIGURE 1. *Graphic representations of proportions presented in Table 1.*

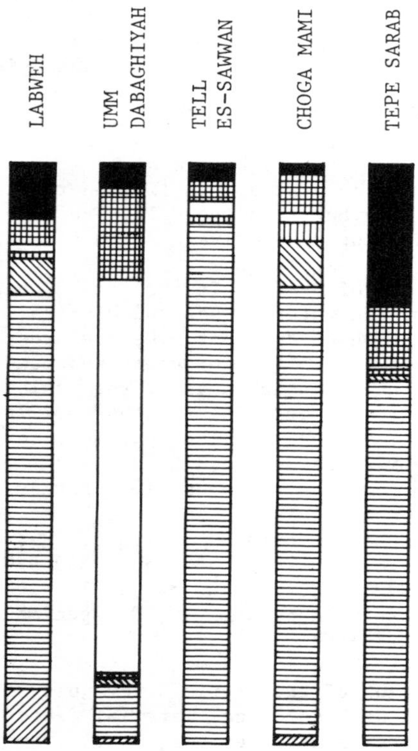

TABLE 1. The percentages of domestic animals and of the major groups of wild animals

	Labweh	Umm Dabaghiyah*	Tell es-Sawwan	Choga Mami	Tepe Sarab**
cattle	9.42	1.03	0.07	1.24	?
sheep/goat	67.26	8.97	88.74	77.71	62.8
pig	6.95	1.04	0.07	7.74	0.2
dog	0.40	0.52	1.40	2.63	1.0
domestic animals	84.03	11.56	90.28	89.32	64.0
onager	1.29	68.35	2.44	1.55	0.01
gazelle	4.76	15.99	4.00	7.89	11.5
other wild animals	9.92	4.10	3.28	1.24	24.5
wild animals	15.97	88.44	9.72	10.68	36.0

* first two seasons only   ** approximate percentages

The wild faunas of the two groups of sites mentioned above are different not only in their relative abundance but also in the nature of the species represented (Tables 1 and 2). At Umm Dabaghiyah, real forest species do not occur at all. The aurochs and the rare badger could have lived also in the forest steppe (they were probably hunted in the foothills of the Jebel Sinjar), and the wild swine, again very rare, has a reasonable habitat in the bush around the salty lakes. At the same time, the most frequently encountered animal, the onager, as well as the gazelle (whose frequency is still higher than that of all domestic species taken together), and the hyaena are typical steppe animals, and their high frequencies point to an environment very similar to that of the Jezira region today.

In the faunas of Labweh, Tell es-Sawwan, Choga Mami, and Tepe Sarab, real forest species also appear, but their frequencies depend heavily upon the location of the sites. For example, at Choga Mami, one single fallow deer bone represents this group; the animal was probably hunted in the Zagros Mountains nearby. The fallow deer is the only forest element present also in the assemblage from Tell es-Sawwan. Its habitat must have been the forest strip along the Tigris River, similar to its habitat in modern Iran (Reed 1965). Not surprisingly there are more remains of forest animals in the Labweh sample since the site is located close to two mountain ranges. Here, the red deer is the most frequent member of the wild fauna, and, in addition, the fallow deer and even the brown bear occur. The aurochs is also rather common, probably hunted in the thin forests of the wide open valley (Liere and Contenson 1964), while the onager is very rare. Forest animals appear in the highest frequencies at Tepe Sarab, located as it is in the middle of the high Zagros. Red deer, fallow deer, roe deer, wild swine, leopard, wild cat, brown bear, badger, and beaver have been identified from this site together with a fair amount of wild sheep, goat, gazelle, and just a few onagers.

In contrast to the wild fauna, the domestic faunas of the five sites are very uniform in their characteristics. 1) At each of the sites, all five domestic species were present. (The only possible exception in this respect is Tepe Sarab where domestic cattle may not have been present.) 2) In the domestic fauna of each site, the caprovines—sheep and goat—are by far the most frequently occurring animals. At the same time, the five sites differ markedly in the frequencies of the other three domestic species. At Labweh, cattle were more frequent than pigs; at Tell es-Sawwan, they were equal; at Choga Mami, pigs were more numerous than cattle; at Sarab, domestic cattle were possibly not present; and at Umm Dabaghiyah, the picture changes almost level by level. The case of the dog is very similar, being almost absent at one site and surpassing the frequency of cattle and pigs at another.

The occurrence of all five Neolithic domestic species in the faunal assemblages of these sites is not surprising since they had appeared together in Southeast Europe much earlier, indeed by the middle of the seventh millennium B.C. (Bökönyi 1974). Although four out of the five animals (dog, sheep, goat, and pig) were domesticated by man in Southwest Asia earlier than in Southeast Europe (with only cattle being domesticated at about the same time in both regions), it is certainly peculiar that a domestic fauna consisting of all five species together occurred in the south of the Balkan Peninsula some five hundred years earlier than in the Middle East (Bökönyi, in press). Nevertheless, one still cannot reach a final conclusion in this matter because the Prepottery Neolithic of Greece had its closest connections with that of west Anatolia, and from the latter region there have not yet been studied in detail any faunas from sites earlier than the seventh millennium.

The high caprovine ratio from each of the five Near Eastern sites discussed here is also not surprising. Caprovines represent 54 to 90 percent of the domestic fauna from the different levels of Umm Dabaghiyah, 80 percent at Labweh, 87 percent at Choga Mami, about 97 percent at Tepe Sarab, and 98 percent at Tell es-Sawwan. Southwest Asia has always been the real homeland of caprovines. Their wild ancestors lived there in great abundance, and they still exist there today. Their domestication first took place in that region, and neolithic man had the opportunity of increasing the size of his herds through local domestication, there being a large wild stock at hand. In fact, such conditions alone were sufficient reason for a species to become dominant in a domestic fauna in prehistoric times. In primitive animal husbandry, the progeny of domestic animals was not sufficient both to maintain (or increase) the herd size and to supply the meat requirements of the people. Therefore those species which had domesticable wild counterparts in a given region were preferred by ancient breeders and consequently became dominant in the domestic fauna. In addition, the environmental conditions of Southwest Asia were quite favorable to both species of caprovines, and there is little wonder, therefore, that they bred well there.

Turning now to the relative frequencies of the two small domestic ruminants, sheep are more frequent than goats at Labweh and Umm Dabaghiyah (sites located in the western part of the region) with the situation being just the opposite for the three other sites. At Labweh, sheep just surpass goats in frequency of occurrence, the ratio being 51 to 49 percent. At Umm Dabaghiyah, sheep dominance is better expressed, the ratio being 66 to 34 percent. At Tell es-Sawwan, the ratio is the same but with goats predominating while at Choga Mami and Tepe Sarab, goats are in an overwhelming majority—88 to 12 and 86 to 14 percent, respectively. These data contradict Flannery's view that in mountainous areas goats were more numerous, while in the plains sheep were dominant in prehistoric times (Hole and Flannery 1967). Flannery's statement is valid for Tepe Sarab and Umm Dabaghiyah and perhaps also for Choga Mami, but it is certainly not true for Labweh and Tell es-Sawwan. When Flannery formed his opinion, he purportedly based it on ecological considerations alone (Flannery 1965). The data from these five sites, however, suggest that the causes are probably rooted in the domestication process itself. It is quite possible that goat domestication was larger in scale in the eastern part of the area and sheep domestication was more important in the western part.

The occurrence of domestic cattle in at least

four of the sites demonstrates how quickly this species spread out of its first center of domestication in the eastern basin of the Mediterranean (including Anatolia) after the middle of the seventh millennium B.C. (Bökönyi 1973a). The main route for this extension initially must have been along the eastern coast of the Mediterranean and into its neighboring areas, a consideration which suggests why domestic cattle appear in the Prepottery Neolithic of Jericho (Nobis 1968) and why they are so frequent at Labweh. At the same time, cattle were much

TABLE 2. Faunal lists (presence/absence)

	Labweh	Umm Dabagh-iyah	Tell-es -Sawwan	Choga Mami	Tepe Sarab
domestic animals					
cattle	+	+	+	+	?
sheep	+	+	+	+	+
goat	+	+	+	+	+
pig	+	+	+	+	+
dog	+	+	+	+	+
wild animals					
aurochs	+	+	+	+	+
wild sheep	−	+	−	−	+
wild goat	−	+	−	−	+
gazelle	+	+	+	+	+
red deer	+	−	−	−	+
fallow deer	+	−	+	+	+
roe deer	−	−	−	−	+
wild swine	+	+	+	+	+
onager	+	+	+	+	+
leopard	−	−	−	−	+
wild cat	−	−	−	−	+
large cat	−	−	−	+	−
badger	−	+	−	−	+
brown bear	+	−	−	−	+
hyaena	−	+	−	−	−
wolf	−	+	−	−	+
fox	+	+	+	−	+
other carnivore	−	+	−	−	+
beaver	−	−	−	−	+
hare	−	+	+	−	+
hedgehog	−	−	−	−	+
birds					
red-billed chough	−	−	−	−	+
european bee-eater	−	−	−	−	+
great bustard	−	−	−	−	+
european crane	−	−	−	−	+
rock partridge	−	−	−	−	+
pallid harrier	−	−	−	−	+
rough-legged buzzard	−	−	−	−	+
golden eagle	−	−	−	−	+
lesser white-footed goose (?)	−	−	−	−	+
bean goose	−	−	−	−	+
unidentified	+	+	+	+	+
other					
tortoise	+	−	−	−	+
belenide	−	−	−	−	+
fishes (unidentified)	−	−	+	−	+

rarer at the other sites, and seem not to have penetrated the Zagros Mountains in the sixth millennium B.C. They had, however, already reached the foothills of the Zagros as is indicated by their occurrence in the Sabz phase of Tepe Sabz (Hole and Flannery 1967) and at Choga Mami. In the sixth millennium B.C., domestic cattle appeared also in Turkestan (Calkin 1970) during the early phase of the Dzeitun Culture. Since wild cattle bones were not found in the sites, one presumes that domestic cattle were imported from some other source. On the basis of the pottery decorations, Brentjes (1971) has suggested that this culture originated in west Anatolia and on the Iranian plateau (Sialk) which, if true, would provide interesting evidence for the radiation of early domestic cattle in the Middle East.

The pig seems to be an animal domesticated earlier than cattle, although this is based only on Flannery's statement that the domestic pigs of Jarmo (ca. 6500 B.C.) were not of local origin but came from another population domesticated elsewhere and at an earlier date (Flannery 1961; Reed 1961). On this basis, the occurrence of domesticated pigs might be expected in all five sites. The comparatively low pig ratio at all of these sites may be related to the dry environment, the pig's preference for wet environments being quite well known.

The dog whose earliest domestication took place in northeast Iraq during the terminal Pleistocene (during the Zarzian period, ca. 12,000 to 11,000 B.C., Turnbull and Reed 1974) had no economic importance at any of these sites although it was already widely spread throughout the Middle East and Europe by the early Neolithic.

And finally it is necessary to say some words about the specialized onager hunting at Umm Dabaghiyah. As mentioned previously, the site is rather unique in this respect and has the largest onager bone sample ever found. Similar specialized hunting has been documented for other animals in the Middle East (wild sheep and goat), in the Carpathian Basin (aurochs), and in the south Ukraine (wild horse) and in each case developed into the local domestication of the hunted species. Therefore the possibility of local domestication must be checked on the onager bone sample from Umm Dabaghiyah.

Demonstration of the existence of local onager domestication can be approached from two directions. One approach involves the study of anatomical changes caused by domestication and the other requires the documentation of changes in the age- and sex-group ratios of the species. The first method is the most reliable but, particularly in equids, anatomical changes seem to appear rather a long time after domestication and, in addition, have yet to be described in detail. As for the second method, on the one hand, sex-group ratios are almost impossible to determine because of the difficulty of determining sex from fragmented bones and loose teeth such as those found in the midden remains of Umm Dabaghiyah. On the other hand, the large tooth sample provides excellent data for the determination of age-group ratios and for demonstrating any changes in those ratios over time.

As expected, anatomical changes resulting from domestication could not be found on the onager bones from Umm Dabaghiyah. As for changes in age-group ratios, the situation is as follows: for the domesticated species (sheep, goat, cattle, and pig) the ratio of immature individuals is high. This situation is quite understandable since man killed such animals for their meat and kept only valuable breeding stock through the winter, feeding them on scarce fodder. At Umm Dabaghiyah, the ratio of immature animals is 43 percent for cattle, 42 percent for caprovines, and 84 percent [!] for pig. This last figure is so high because pigs are multiparous and man need keep a far smaller breeding stock to maintain the population at a given level. For the wild species, the ratio is low with about 21 percent of the aurochs and 17 percent of the gazelle being young. In this respect, the onager sample resembles those of the wild species, revealing as it does a low ratio of immature individuals--21 percent. This fact suggests that the local domestication of onager cannot be demonstrated for Umm Dabaghiyah.

In summary, the study of the faunal remains from these five Neolithic sites suggests that primarily the environment (including geography, climate, soil, flora, and fauna) dictated the nature of the animals hunted or kept by the inhabitants, and that cultural factors were of only secondary importance. For the wild fraction, the environment was clearly the limiting factor because man could only hunt those species which preferred the environmental zones existing in the vicinity of the settlements. Hunters exploited all environmental zones within a certain radius of the settlement and sometimes even covered larger distances in order to kill particularly choice prey. For example, the hunters of Umm Dabaghiyah seemed to have traveled for about three days on foot to the Jebel Sinjar to hunt the aurochs, wild sheep, and goat, and possibly even badger. Specialized hunting occurred at Umm Dabaghiyah (the onager) and perhaps also at Tepe Sarab (the wild goat). The first case seems not to have led to the domestication of the hunted species while the second was a continuation of a local tradition started in the ninth millennium B.C.

The environment seems to have exercised a certain amount of influence on the domestic fauna as well, a fact that is suggested by the high ratio of caprovines in the domestic animal remains from each site. Here the influence of the environment was twofold, providing domesticable wild stock and grazing areas preferred by this type of domestic animal. Domesticable wild caprovines seem to have been far more abundant than aurochs or wild pig in the area, and it is therefore little wonder that their domesticated descendants are more frequently represented than other domestic animals. In addition, the dry environment of the area, similar to that of today, was one much more favorable to sheep and goat raising than to supporting cattle and particularly pigs.

The environment had less of an effect on determining the importance of husbandry relative to hunting, although in extreme cases such as that of Umm Dabaghiyah, it certainly left its mark. In this domain, the primary influence was probably cultural as indeed was specialized hunting a primarily cultural response. Such clearly marked cultural differences in the fauna such as those evident in the Neolithic of Europe are not found here, however, although they may appear in later phases of the Neolithic. Unfortunately not enough faunal material from later sites in the Middle East has been studied.

References

Bökönyi, S.
   1973a   "Some problems of animal domestication in the Middle East," in J. Matolcsi, editor, *Domestikationsforschung und Geschichte der Haustiere*, pp. 69-75. Budapest.
   1973b   "Stock breeding," in D.R. Theocharis, *Neolithic Greece*, pp. 165-178, Athens.
   1973c   "The fauna of Umm Dabaghiyah: a preliminary report," *Iraq*, vol. 35, pp. 9-11.
   1974   *History of Domestic Mammals in Central and Eastern Europe*. Budapest.
   1976   "The vertebrate fauna of the neolithic site at Anzabegovo," in M. Gimbutas, "Neolithic Macedonia," *Monumenta Archaeologica I*, pp. 313-363. Los Angeles.
   1977   *The Animal Remains from Four Sites in the Kermanshah Valley, Iran: Asiab, Sarab, Dehsavar and Siahbid; the faunal evolution, environmental changes and development of animal husbandry, VIII-III millennia B.C.* British Archaeological Reports Supplementary Series, no. 34. Oxford.
   in press   "Die Herkunft bzw. Herausgestaltung der neolithischen Haustierfauna Südosteuropas und ihre Beziehungen zu Südwestasien."

Boessneck, J.
   1962   "Die Tierreste aus der Argissa Magula vom präkeramischen Neolithikum bis zur mittleren Bronzezeit," in V. Milojcić, J. Boessneck, and M. Hopf, *Die deutschen Ausgrabungen auf der Argissa Magula in Thessalien I*, pp. 27-99. Bonn.

Braidwood, R.J.
   1960   "Seeking the world's first farmers in Persian Kurdistan: a full-scale investigation of prehistoric sites near Kermanshah," *The Illustrated London News*, October 22, pp. 695-697.
   1969   "The earliest village communities of Southwestern Asia reconsidered," *Atti del VI Congresso Internazionale delle Scienze Preistoriche Protoistoriche I*, pp. 115-126.

Brentjes, B.
   1971   "Die Entwicklung im Vorderen Orient vom 9-4. Jahrtausend," in F. Schlette, editor, *Evolution und Revolution im Alten Orient und in Europa*, pp. 23-37. Berlin.

Calkin, V.I.
   1970   "Drevnejsie domasnie zivotnye Srednej Azii (The most ancient domestic animals of Middle Asia) Soobscenie I," *Bjulleten' Moskovskogo Obscestva Ispytatelej Prirody - Otdel Biologiceskij*, vol. 75, pp. 145-159.

Flannery, K.V.
   1961   "Skeletal and radiocarbon evidence for the origins of pig domestication," M.A. thesis, University of Chicago.
   1965   "The ecology of early food production in Mesopotamia," *Science*, vol. 147, pp. 1247-1256.

Gejvall, N.-G.
   1969   *Lerna, Volume I: The Fauna*. Princeton.

Higgs, E.S.
   1962   "The fauna of the early neolithic site at Nea Nikomedeia, Greek Macedonia," *Proceedings of the Prehistoric Society*, vol. 28, pp. 271-274.

Hole, F. and K.V. Flannery
   1967   "The prehistory of Southwestern Iran: a preliminary report," *Proceedings of the Prehistoric Society*, vol. 33, pp. 147-264.

Jarman, M.R. and H.N. Jarman
   1968   "The fauna and economy of early neolithic Knossos," *Annals of the British School of Archaeology at Athens*, vol. 63, pp. 241-264.

Kirkbride, D.
   1972   "Umm Dabaghiyah 1971: a preliminary report," *Iraq*, vol. 34, pp. 3-15.

Liere, W.J. van and H. de Contenson
   1964   "Holocene environment and early settlement in the Levant," *Annales Archéologique de Syrie*, vol. 14, pp. 125-128.

Nobis, G.
   1968   "Säugetiere in der Umwelt frühmenschlicher Kulturen," in M. Claus, W. Haarnagel, and K. Raddatz, *Studien zur europäischen Vor- und Frühgeschichte*, pp. 413-430. Neumünster.

Reed, C.A.
   1961   "Osteological evidences for prehistoric domestication in Southwestern Asia," *Zeitschrift für Tierzüchtung und Züchtungsbiologie*, vol. 76, pp. 31-38.
   1965   "Imperial Sasanian hunting of pig and fallow deer, and problems of survival of these animals today in Iran," *Postilla* (Peabody Museum of Natural History, Yale University), vol. 92, pp. 1-23.

Turnbull, P.F. and C.A. Reed
   1974   *The fauna from the terminal Pleistocene of Palegawra Cave, a Zarzian occupation site in northeastern Iraq.* Fieldiana: Anthropology, vol. 63, no. 3, pp. 81-146. Chicago.

RODENTS AND THE ARCHAEOLOGICAL PALEOENVIRONMENT:
CONSIDERATIONS, PROBLEMS, AND THE FUTURE

Richard W. Redding, Museum of Zoology
University of Michigan

One of the problems that zooarchaeologists and paleontologists have attempted to solve with data provided by faunal remains is the reconstruction of environments around a site at the time of deposition. The success of such studies and the detail of the reconstruction have been variable. One character common to most of these studies is that rodents are utilized as the primary group, upon which the reconstruction is based, or are cited in support. Rodents hold the greatest potential for archaeologists for monitoring paleoenvironments because they are more sensitive to changes in the local environments of an archaeological site than are larger mammals. Yet most rodents are not adapted to such restricted niches as for instance, the gastropods, so that they are found in numerous communities.

The preferred environments of rodent taxa utilized in the published reconstructions have been characterized in such general terms as "a dry form," "representative of damp conditions" (Bate 1937), and "boreal" (Chaline 1977) or have been described in such probabilistic phrases as "typically burrows in deep wet soil" (Turnbull and Reed 1974), "an indicator of moist meadow and parkland" (Gruhn 1961) and "lives in damp grasslands" (Chaline 1977). General descriptive terms provide a minimum of information. Probabilistic statements are based upon contemporary subjectively observed associations between rodent taxa and the environment. They are analogical rather than explanatory in nature. One should not be led into thinking, as in Chaline (1977, p. 48), that much is known of the tolerances of or the optimal conditions for any given rodent species. The factors that control the distribution within the range of occurrence for most rodents are unknown or poorly delineated. Rodents provide information about the environment in terms of these factors by their presence in an archaeological site. If the factors controlling the local distribution of a rodent species can be determined, then any study of paleoenvironments using that species will be more detailed and precise.

Some of the factors that control the local distribution of rodent species and the relevance of these factors to the monitoring of paleoenvironments are considered below. This is followed by an examination of the problems involved in using rodents to monitor paleoenvironments. Finally, as an exercise, preliminary data on the ecology of two rodents from the Susiana Plain in southwestern Iran will be utilized to monitor the environments of Tepe Ali Kosh on the nearby Deh Luran Plain.

Potential Controlling Factors

Those ecological parameters of a living rodent species that are important in paleoenvironmental interpretations are the factors that control the occurrence of that species at a given locality within its range. Factors affecting the density of a species, however, are meaningless since data on paleo-densities are unattainable from archaeological samples. Numerous factors may control the local occurrence of a rodent species, but some of them provide information that cannot be used to monitor paleoenvironments either because they are not environmental in nature or because they are negative in character (i.e., only explain the absence).

Dispersal and population crashes are two factors that are negative in character. An area may be suitable in all respects for a species but be uninhabited because there is a barrier to dispersal or the population has been removed by epidemic disease or heavy parasitism. If an area is suitable and a species disperses into it or the population in it is unaffected by excessive disease/parasitism, then these factors are trivial in explaining the presence of the species. Such factors can explain only the absence of a species from the area around a site. The condition of absence is not determinable from an archaeological sample.

There are numerous factors that could be considered as physiological in nature, with temperature tolerance and water requirements being the major ones. Rodents utilize behavioral adaptations and microenvironments to avoid environmental extremes of temperature and aridity (e.g., Kennerly 1964; Hoover et al. 1977). This allows rodents to withstand greater variation in physical parameters than one would predict on the basis of the physiological tolerances of the species (see Heller and Gates 1971). It is unlikely that physiological factors are a variable controlling the distribution of most rodents (e.g., Getz 1961a, 1961b, 1971; Brower and Cade 1966; Heller 1971; Miller and Getz 1972; Master 1977; however, see Grubb 1974, on Mus musculus; Hoover et al. 1977).

The presence of a particular food resource or group of resources is a factor that controls the distribution of most rodents. It may be the primary factor or one of two affecting local distribution (Getz 1961a, 1961b; Reig 1970; M'Closkey and Lajoie 1975; Merritt 1974). For those species in which it can be demonstrated that food is a controlling factor, the presence of the particular food resources utilized is the information provided by the species when it is found at a site. But the local distributions of prey species are in turn controlled by one or more factors. Given a herbivorous rodent whose presence at a locality is explained by the presence of a group of food resources which are themselves dependent upon a set of physical-biotic factors, the presence of that rodent at an archaeological site may be utilized to infer the presence of the set of physical-biotic conditions. Getz (1961b) found that the local distribution of Blarina brevicauda, the short-tailed shrew, is controlled by the density of large invertebrates and the level of substrate humidity. Possibly, the level of humidity is the primary factor controlling the density of large invertebrates which then in turn controls the local distribution of B. brevicauda.

Competition is a factor which must be considered but is rarely tested (however, see Grant 1972; Master 1977). For a given species A, it may appear as if food resources are the primary factor controlling its distribution. However, species A may be capable of utilizing a second group of resources but does not because, in those areas in which the second group is found, it is excluded by competition with species B. This factor is apparently negative in nature. But, consider the case in which A is found at an archaeological site and the controlling factor has been assumed to be a set of food resources. In reality A is in competition with B for a second group of food in a second community. If species B was absent in such an area it could be inhabited by A (e.g., Master 1977). Is the presence of A at the archaeological site an indicator of a set of physical-biotic conditions supporting a group of food resources or of the absence of B?

If competition is the primary factor controlling the distribution of a species then that species cannot be used to monitor paleoenvironments. Competition is not environmental in nature, although it may have an environmental component. If competition is one of two or more factors controlling the local distribution of a species then, on one condition, the species may be utilized to monitor paleoenvironments. The condition is that in the site, at the same level as remains of the species under consideration, all species known to affect its distribution, based on modern ecological studies, be present. In such a case the information provided by the presence of the species at the site is in terms of the remaining controlling factor(s).

Predation can function to control the local distribution of a species (Ordzie 1976). It has been accepted or rejected as a controlling factor in most studies without systematic testing (Getz 1961a; Heller 1971; Miller and Getz 1972) or has been ignored (e.g., Getz 1961b). There is disagreement over the potential effect of predation on mammal populations (Vaughan 1978). There is little doubt that predators can affect the density of rodent populations but the ability of predators to control local distributions is debatable. Miller and Getz (1972) considered predation to be the primary factor controlling the local distribution of Clethrionomys gapperi, the red back vole, but expressed it in terms of density of debris and cover that would act as protection against predation.

The presence of suitable shelters or sheltering conditions may act to control local distributions. Vegetation may provide the shelter in which case the controlling factor may be expressed in terms of percent of shrubby vegetation (Brower and Cade 1966) or density of vegetative debris on the surface (Miller and Getz 1972). Burrows may provide the shelter, in which case the controlling factor may be expressed in terms of the physical structure of the soil (Miller 1964; Best 1973; Hoover et al. 1977) or in terms of pre-prepared burrows (Merritt 1974).

Physiological tolerances, food resources, competition, predation, and sheltering conditions are factors that may affect the local distribution of rodent species and which are of importance in using rodent species to monitor paleoenvironments. Food resources and competition seem to be the most important factors. Competition is not environmental in nature but does affect the usefulness of a species for monitoring paleoenvironments. Food resources as a controlling factor may be expressed in terms of the physical-biotic factors that control that prey if the data are available in the literature. Predation and sheltering conditions (in part) appear to be interrelated and to be of secondary importance. Sheltering conditions, in terms of soil structure, appear to be important for fossorial (burrowing) species. Physiological tolerances may ultimately determine the range of a species but are probably not important in controlling local distributions. No factor, however, should be eliminated without adequate testing.

Using Rodents to Monitor Paleoenvironments

Four conditions must be satisfied in order to utilize the rodent remains from an archaeological site to monitor the paleoenvironment. The one most frequently ignored or uncritically accepted (e.g., Chaline 1977, p. 48) is that the species utilized has not significantly changed habitat since its inclusion in the site. A second condition is that the species utilized not be domesticated or managed. Third, the material utilized for the paleoenvironmental study must be drawn from the local environments. Finally, none of the material may be intrusive from later deposits.

The assumption made by many workers that the habitat of a species has not changed has been criticized previously (McCown 1961). There are techniques which may be employed to satisfy this condition. One of these, applicable to post-Pleistocene sites, is to demonstrate that the relative composition of the types of communities available for exploitation by the species has not changed (i.e., extensive areas in the region have not been occupied by new types of communities and no community types have been lost in the region). This must be accompanied by evidence that the rodent community has not changed (i.e., no rodent species not in the modern range of the species utilized in the paleoenvironmental study have been introduced and none lost). This method of satisfying the first condition is not as powerful as some others because it is indirect. It is, however, adequate and simple.

The second condition is commonly recognized. A species that is domesticated or managed loses all meaning in paleoenvironmental studies.

The condition that the sample from the site be representative of the fauna around the site is dependent upon the delineation of the mechanism of introduction. One must be suspicious of rodent taxa introduced into a site as a food resource of humans. Such a mechanism may introduce species that come from a considerable distance from the site and in numbers out of proportion to their occurrence in the rodent community. Rodents, except for domesticated species, are unlikely to represent a human food resource. The only Middle Eastern report known to the author which maintains that small rodents were a human food resource is that on Palegawra (Turnbull and Reed 1974). The conclusion that the rodents Spalax leucodon and Arvicola terrestris were a human food resource is based on their presence in a cave which is a habitat far removed from their expected habitat. The authors maintain that the rodents must have been brought into the site by humans as a food resource. An alternate explanation is found in the Palegawra bird list which includes Tyto alba, the barn owl, and Falco tinnunculus, the kestrel, both of which will inhabit caves and feed on rodents. T. alba is known to feed on S. leucodon and A. terrestris (Dor 1947).

At most sites rodents are probably introduced by birds of prey roosting at the site, by domestic and wild carnivorous mammals, or by the inhabitants occasionally or accidentally introducing them, and the rodents themselves attempting colonization. None of these mechanisms of introduction would result in rodents being brought in from outside the local environments of the site. Most small to medium raptors, in North America, forage in an area of 0.10 to 5.6 km2 (Craighead and Craighead 1956). The red fox, Vulpes vulpes, in North America, has a home range of 1.6 to 6.0 km^2 (Fox 1975). None of these mechanisms of introduction produce an unbiased sample of the rodent fauna of the area around the site. But an unbiased sample is only important when comparing absolute numbers or computing ratios. It is not important when dealing only with presence of a species.

The condition that none of the material utilized in the study be intrusive is not difficult to satisfy if the site has been carefully excavated. Criteria for distinguishing intrusive material have been discussed elsewhere (Hole et al. 1969; Redding n.d.). The three criteria most useful and reliable are depth within the deposit, association with burrow fill or other disturbed areas, and comparison of discoloration and degree of mineralization with material from the same level known not to be intrusive.

An Example

A study begun in 1973 of Tatera indica, the Indian gerbil, and Meriones crassus, Sundevall's jird, on the Susiana Plain in southwestern Iran, was designed to determine plant community associations of the two species and to obtain qualitative information on the food resources utilized. The data are incomplete but are briefly summarized below. This is not intended as a complete study to define the controlling factors for these two rodents. No quantitative data on competition and predation for either species were collected.

The Susiana Plain, and nearby Deh Luran Plain, may be characterized as semiarid steppe. They receive 200 to 399 mm of rain each year falling mainly between late November and late April (Beaumont et al. 1976).

T. indica on the Susiana Plain utilize the seeds, stems, rhizomes, and leaves of grasses (Graminae) as well as several species of insects and, rarely, seeds of Zizyphus spina-christi, the jujube or Kanar (Rhamnaceae). T. indica is restricted to those areas that support grasses for most of the year. On the Susiana Plain this implies an association with a river, stream, swamp, or irrigation canal. The leguminous component of the flora associated with T. indica is composed predominately of species producing dehiscent (those which split open to discharge the seeds) seedpods. This may be another indication of moist soil conditions that persist for a major part of the year. This account of the food habits of T. indica is substantially supported in the literature (Blanford 1888; Prasad 1954).

M. crassus on the Susiana Plain primarily utilize seeds of legumes producing indehiscent (those which do not split open but fall to the ground retaining the seeds) seedpods. They also utilize green parts of plants during the growth season and insects. On the Susiana Plain, communities with a predominately indehiscent leguminous component are restricted to marginal agricultural land that is not irrigated or gravel ridges found on and around the plain. This summary is only partially confirmed by literature on M. crassus, most of which is anecdotal (Vesey-Fitzgerald 1953; Lewis et al. 1965; Lay 1967).

Let us assume, in order to continue our example, that food resources are the primary factor controlling the distribution of these two rodents. Observational data suggest that competition is not important (Redding, unpublished data). There remains the possibility that the distribution of T. indica may be controlled by the presence of free water.

In 1963-64 the site of Tepe Ali Kosh was excavated on the Deh Luran Plain in southwestern Iran (Hole et al. 1969). Remains of T. indica and/or M. crassus were found in all six zones that represent three phases spanning the time between 8200 and 6100 B.C. However, before the two rodents are utilized to examine the paleo-environments around Ali Kosh the four conditions noted above must be satisfied.

A study of the geomorphology of the Deh Luran Plain by Kirkby (1977) provides data that indicates that the relative size of environmental zones in the region has varied but that no zone related to either T. indica or M. crassus has disappeared or been severely restricted. Kirkby explains those shifts in relative sizes of environmental zones that have occurred as the result of changes in the hydrology of the two river systems on the plain. There appears to have been no major change in precipitation on the plain in the last 9000 years. (This last statement, however, is not entirely supported by the work of Nützel 1976.) The rodent community on the Deh Luran Plain at present includes Nesokia indica, Mus musculus, and Gerbillus nanus as well as T. indica and M. crassus. All of these species, except G. nanus, are known from archaeological sites on the Deh Luran Plain which cover the time period from 8200 to 1800 B.C. G. nanus, on the Deh Luran Plain is restricted to areas of soft silt on the flood plains of the rivers (Redding, unpublished data). Their absence from the small sample of archaeological material from the area does not suggest their absence from the rodent fauna of the plain prehistorically. No rodent species have been found in the archaeological sample that are not members of the modern fauna of the plain.

The evidence on the stability of the environment and rodent fauna of southwestern Iran during the last 10,000 years presented above suggests that the assumption that T. indica and M. crassus have not shifted habitat is not unreasonable. This satisfies the first condition.

The condition that the two rodent species were neither domesticated nor managed can be satisfied by noting that the small numbers of both in the samples make it unlikely that either was maintained or manipulated. Further, there is no evidence on any of the rodent elements of butchering, burning, or use for any purpose.

The mechanism of introduction of the rodents into the site of Ali Kosh is unknown. While they were probably not a food resource for humans (see above), they probably were for wild carnivorous mammals. The wild cat, Felis catus, the red fox, Vulpes vulpes, and a weasel, Mustela nivalis (?) were present in the area as indicated by their remains in the site. The rodents were probably not introduced from outside the local environments of the site.

The fourth condition, that the sample not be contaminated by intrusive material, can be satisfied. Hole et al. (1969) state that recent intrusive sediments in burrows were recognized and removed and only heavily discolored elements were utilized in their analysis.

The distribution of remains of T. indica and M. crassus within the site of Ali Kosh is presented in Table 1. From the distribution of T. indica it is clear that during all phases the local area around Ali Kosh supported stands of grasses that persisted throughout most of the year. The leguminous component of the flora in these areas consisted of species producing dehiscent pods. Relative to other communities

TABLE 1

Counts by phase and zone of mandibles and maxillae of Tatera indica and Meriones crassus from Tepe Ali Kosh (dates from Hole 1977)

Phase	Zone	T. indica	M. crassus
Mohammad Jaffar	A1	24	8
(c.6400-6100 B.C.)	A2	31	8
Ali Kosh	B1	12	-
(c.7200-6400 B.C.)	B2	129	-
Bus Mordeh	C1	75	1
(c.8200-7200 B.C.)	C2	31	-

on the plain, such areas had high soil moisture that was the result of local perennial sources of water. During zone C1 of the Bus Mordeh phase and both zones of the Mohammad Jaffar phase the presence of M. crassus is explained by the occurrence around the site of areas supporting indehiscent pod-producing legumes. These areas may have been farmed but were not irrigated nor did they benefit from nearby perennial bodies of water. M. crassus is restricted, with the exception of a single element, to the last phase at Ali Kosh. It is tempting to suggest that this distribution is the result of environmental changes in area around the site. These data suggest that the factors controlling the distribution of M. crassus were at a favorable level only on the margin of the area within the radius of ordinary human use or in small localized areas within it during the Bus Mordeh and Ali Kosh phases but became more common during the Mohammad Jaffar phase. Thus, to generalize, it appears that the local environment of Ali Kosh became drier during the Mohammad Jaffar phase.

Since the excavations at Ali Kosh resulted in the recovery of seeds and other plant materials, the data provided by the rodents can be tested. The report of the botanical remains is excellent but since it is preliminary it does not include counts by phase of many of the species (Helbaek 1969). It supports all the statements and predictions made above on the paleoenvironment of Ali Kosh. A single exception is the predicted increase of indehiscent legumes in the Mohammad Jaffar levels which cannot be tested because Helbaek combines the legumes.

Conclusion

Rodents have the potential to provide paleo-environmental information not attainable from other animal groups. Plant remains are potentially as useful but pass through a cultural filter before their deposition and are infrequently preserved. The potential of rodents has not been realized because their preferred habitats which are utilized in the reconstructions have been described in only the most general terms or by analogy. Rodents from archaeological sites provide paleoenvironmental information in terms of those factors that control their local distribution. Food resources and competition are probably the most commonly operating controlling factors. But physiology, predation, and sheltering conditions play a role in determining the local distribution of some species. For most rodent species the controlling factors are unknown or poorly defined. To maximize the utility of rodents, the necessary information must be solicited from mammalian ecologists or collected by the zooarchaeologist. This is yet another reason for the inclusion of a zoologist or zooarchaeologist on every field project.

References

Bate, D.M.
- 1937 "The fossil fauna of the Wadi el-Mughara caves," in D.A.E. Garrod and D.M. Bate, *The Stone Age of Mount Carmel*, vol. 1, pt. 2, pp. 136-240. Oxford.

Beaumont, P., G.H. Blake, and J.M. Wagstaff
- 1976 *The Middle East: A Geographical Study*. New York.

Best, T.L.
- 1973 "Ecological separation of three genera of pocket gophers (Geomyidae)," *Ecology*, vol. 54, pp. 1311-19.

Blanford, W.T.
- 1888-1891 *Mammalia*. The Fauna of British India, including Ceylon and Burma, vols. 5 and 20. London.

Brower, J.E. and T.J. Cade
- 1966 "Ecology and physiology of Napaeozapus insignis (Miller) and other woodland mice," *Ecology*, vol. 47, pp. 46-63.

Chaline, J.
- 1977 "Rodents, evolution and prehistory," *Endeavor*, n.s., vol. 1, no. 2, pp. 44-51.

Craighead, J.R. and F.C. Craighead
- 1956 *Hawks, Owls and Wildlife*. Harrisburg.

Dor, M.
- 1947 "Observations sur les micromammifères trouvés dans les pelotes de la chouette effrayé (Tyto alba) en Palestine," *Mammalia*, vol. 1, pp. 50-54.

Fox, M.W.
- 1975 *The Wild Canids: Their Systematics, Behavioral Ecology and Evolution*. New York.

Getz, L.L.
- 1961a "Factors influencing the local distribution of Microtus and Synaptomys in southern Michigan," *Ecology*, vol. 42, pp. 110-119.
- 1961b "Factors influencing the local distribution of shrews," *American Midland Naturalist*, vol. 65, no. 1, pp. 67-88.
- 1971 "Microclimate, vegetation cover, and local distribution of the meadow vole," *Transactions of the Illinois Academy of Science*, vol. 64, no. 1, pp. 9-21.

Grant, P.R.
- 1972 "Interspecific competition among rodents," *Annual Review of Systematics and Ecology*, vol. 3, pp. 79-106.

Grubb, D.
- 1974 "Desert rodents and desert agriculture: the ecological and physiological relationships of a desert rodent community to native and modified environments." Ph.D. thesis, Univ. of California, Irvine.

Gruhn, R.
- 1961 *The Archeology of Wilson Butte Cave, Southcentral Idaho*. Occasional Papers of the Idaho State Museum, no. 6.

Helbaek, H.
- 1969 "Plant collecting, dry-farming, and irrigation agriculture in prehistoric Deh Luran," in F. Hole, K.V. Flannery, and J.A. Neely, *Prehistory and Human Ecology of the Deh Luran Plain*. Memoirs of the Museum of Anthropology, University of Michigan, no. 1, pp. 383-426. Ann Arbor.

Heller, C.H.
- 1971 "Altitudinal zonation of chipmunks (Eutamias): interspecific aggression," *Ecology*, vol. 52, pp. 312-313.

Heller, C.H. and D.M. Gates
- 1971 "Altitudinal zonation of chipmunks (Eutamias): energy budgets," *Ecology*, vol. 52, pp. 424-433.

Hole, F.
- 1977 *Studies in the Archeological History of the Deh Luran Plain: The Excavation of Chagha Sefid*. Memoirs of the Museum of Anthropology, University of Michigan, no. 9. Ann Arbor.

Hole, F., K.V. Flannery, and J.A. Neely
- 1969 *Prehistory and Human Ecology of the Deh Luran Plain*. Memoirs of the Museum of Anthropology, University of Michigan, no. 1. Ann Arbor.

Hoover, K.D., W.G. Whitford, and P. Flavill
- 1977 "Factors influencing the distribution of two species of Perognathus," *Ecology*, vol. 58, pp. 877-884.

Kennerly, T.E., Jr.
- 1964 "Microenvironmental conditions of the pocket gopher burrow," *Texas Journal of Science*, vol. 16, pp. 395-441.

Kirkby, M.J.
- 1977 "Land and water resources of the Deh Luran Plain, Khuzistan, Iran," in F. Hole, *Studies in the Archeological History of the Deh Luran Plain: The Excavation of Chagha Sefid*. Memoirs of the Museum of Anthropology, University of Michigan, no. 9, pp. 251-288. Ann Arbor.

Lay, D.M.
- 1967 *A Study of the Mammals of Iran*. Fieldiana: Zoology, vol. 54.

Lewis, R.E., J.H. Lewis, and D.L. Harrison
- 1965 "On a collection of mammals from northern Saudi Arabia," *Proceedings of the Zoological Society of London*, vol. 144, no. 1, pp. 61-74.

Master, L.L.
- 1977 "The effect of interspecific competition on habitat utilization by two species of Peromyscus," Ph.D. thesis, University of Michigan.

McCown, T.D.
- 1961 "Animals, climate, and palaeolithic man," *Kroeber Anthropological Society Papers*, vol. 25, pp. 221-232.

M'Closkey, R.T. and D.T. Lajoie
- 1975 "Determinants of local distribution and abundance in white-footed mice," *Ecology*, vol. 56, pp. 476-472.

Merritt, J.F.
- 1974 "Factors influencing the local distribution of Peromyscus californicus in northern California," *Journal of Mammalogy*, vol. 55, pp. 102-114.

Miller, D.H. and L.L. Getz
1972 "Factors influencing the local distribution of the red back vole, Clethrionomys gapperi, in New England," *Occasional Papers, University of Connecticut, Biological Sciences Series,* vol. 2, no. 9, pp. 115-138.

Miller, R.S.
1964 "Ecology and distribution of pocket gophers (Geomyidae) in Colorado," *Ecology,* vol. 45, pp. 256-272.

Nützel, W.
1976 "The climate changes of Mesopotamia and bordering areas: 14,000 to 2000 B.C.," *Sumer,* vol. 32, pp. 11-24.

Ordzie, C.
1976 "The role of a predator in habitat selection by Fundulus heteroclitus, a brackishwater killifish," Paper read at the annual meeting of the Animal Behavior Society, 20-25 June, 1976, at the University of Colorado, Boulder.

Prasad, M.R.N.
1954 "Food of the Indian gerbil, Tatera indica cuvieri (Waterhouse)," *Journal of the Bombay Natural History Society,* vol. 52, pp. 321-325.

Redding, R.W.
n.d. The fauna of Tepe Farukhabad. Unpublished manuscript, 65 pp.

Reig, O.A.
1970 "Ecological notes on the fossorial octodont rodent Spalacopus cyanus (Molina)," *Journal of Mammalogy,* vol. 51, pp. 592-601.

Turnbull, P. and C.A. Reed
1974 *The Fauna from the Terminal Pleistocene of Palegawra Cave.* Fieldiana: Anthropology, vol. 63, no. 3, pp. 31-146.

Vaughan, T.A.
1978 *Mammalogy.* 2nd edition. Philadelphia.

Vesey-Fitzgerald, D.
1953 "Notes on some rodents from Saudi Arabia and Kuwait," *Journal of the Bombay Natural History Society,* vol. 51, pp. 424-428.

DIFFERENTIATION BETWEEN THE BONES OF CAPRINES FROM DIFFERENT ECOSYSTEMS IN IRAN
BY THE ANALYSIS OF OSTEOLOGICAL MICROSTRUCTURE AND CHEMICAL COMPOSITION

Melinda A. Zeder, Museum of Anthropology
University of Michigan

Introduction

Animal exploitation has been a major component of Near Eastern cultural systems throughout all periods of development. The analysis of faunal material is, then, particularly useful in studies of the development and operation of these systems. To date, zooarchaeology has made its primary contributions to the reconstruction of environments and subsistence economies. In particular, the domestication of bovids and equids and the cultural implications of domestication have been focal points of Near Eastern zooarchaeological research.

In order to approach problems of this nature with faunal material it has been necessary to establish criteria to discriminate among the bones of different Near Eastern taxa, especially between those of wild and domestic species. These osteological distinctions have relied upon observable and measurable differences in the gross morphology of bone (Gromova 1953; Boessneck et al. 1964; Degerbøl 1963; Jewell 1963; Flannery 1961; 1969; Reed 1960; Bökönyi 1973). Although great strides have been taken in the establishment of these distinctions, there are two serious drawbacks to this approach. First, these techniques of identification can be used only on certain parts of specific bones. Failure to retrieve these bones or the blow of an ancient butchering tool or modern pick can seriously hamper the zooarchaeologist in his identification of animal taxa represented. The second drawback is that these techniques of identification rely upon osteological characteristics that result from genotypic changes in animal populations. Such changes often require many generations to develop to a significant degree. The time element involved in the development of genotypic morphological changes in the bones of domestic populations has severely hampered those trying to identify and characterize processes of domestication.

Reconstruction of environments and subsistence economies, however, are but two of the many areas in which zooarchaeological research may aid in the study of culture history and process in the Near East. Some of these other areas include the development of different strategies of hunting or herding; the use of animals in local and regional exchange; class differences in access to meat resources; and the origin of full-time herding specialists, nomads, and their interaction with, and impact on, other groups. All of these areas are particularly relevant to the study of complex societies in the Near East and especially early states, a study to which zooarchaeology has made relatively little contribution.

Before such problems can be approached, however, new techniques of osteological characterization must be developed which allow finer levels of discrimination among the bones of animals. We must be able to differentiate between animals raised in different environments and under contrasting herd strategies. The establishment of criteria for identification of this nature cannot be approached on the taxonomic grounds traditionally employed. Rather, it must be approached in terms of the animal's participation within an ecosystem. Such participation includes the animal's interaction with other animals, especially man, as well as with its environment. An ecological approach would allow us to postulate why certain osteological characteristics might develop as a result of this participation. From this postulation we would be in a position to predict osteological discriminators among animals that participated in different ecosystems. We should then be able to predict where and how to look for these differences.

As this is a more subtle type of discrimination than taxonomic identification, it may require examination of more subtle characteristics of bone morphology. Rather than looking for differences in gross morphology, it might be more profitable to look for differences in bone microstructure and chemical composition, factors which are likely to be more responsive to ecosystem participation. Some attention has already been given to the establishment of osteological differences between wild and domestic animals through examination of bone microstructure. An early pioneer in this line of research was Bökönyi et al. (1965) who documented differences in the thickness of trabeculae of the bones of wild and domestic cattle. Other innovators in this research were Drew, Perkins, and Daly (1971; see also MASCA 1970; 1973) who examined the orientation of apatite crystals in bone, as well as trabecular thickness and shape, in an effort to establish differences in the bones of wild and domestic caprines.

Examination of bone microstructure and chemical composition potentially avoids the two drawbacks of identification based on gross morphology of bone. Theoretically, identification based on such characteristics can be applied to almost any bone in any condition. In addition, changes in microstructure and composition may be the result of phenotypic adaptations induced by interaction with the environment. If these changes are phenotypic in origin they should appear within one generation of animals participating in a particular ecosystem.

The present study was undertaken as a preliminary step toward the establishment of difference among the bones of wild and domestic sheep as a function of ecosystem participation. A sample of twelve domestic sheep was purchased in 1973 while the author was conducting fieldwork in Iran. The domestic sample consists of four animals from each of three ecological groupings (Fig. 1).[1] The first group consists of lowland sedentary sheep from herds in the vicinity of the Deh Luran and Susiana plains. The second group is comprised of lowland-lowland transhumant sheep herded from these plains to the mountains north and eastward, and from the mountains in the Korammabad area to more lowland areas in the south. The third group is made up of highland-highland transhumant sheep from the Khorammabad area, herded to highland areas in the southwest. Four wild specimens were loaned by the Field Museum of Natural History of Chicago from a large collection of Iranian wild sheep collected by Douglas Lay.

The goal of this study was to postulate why and where osteological differences among these animals could be expected, to explore techniques which might be used to detect and measure these differences, and to assess the value of the continuation of such research. Therefore, this study is best viewed as a pilot project. Until a statistically significant sample is examined, none of the osteological differences presented here can be used with confidence to discriminate between these groups of sheep.

Two factors which may contribute to osteological differences among these groups of animals were examined: activity and diet. The first proposition considered was that structural modifications in bone might result from stresses placed on bone as a consequence of different levels and types of muscular activity of animals with contrasting herd adaptations. It was decided to limit the examination of the bones of the Iranian sheep to osteological microstructure. This decision was made because of the possible advantages of the use of micro- over macrostructural characteristics when looking for differences of this nature, and because it was felt that even if the latter existed, sample size mitigated against making any concrete statements about macrostructural discriminators, whereas there are more measurable variables per specimen using microstructural discriminators. Two aspects of the microstructure of bones were examined: crystalline orientation and trabecular thickness. In both aspects, the work of Drew, Perkins, and Daly was reexamined and expanded upon.

The second proposition was that nutritional intake of animals living in different environments may cause detectable differences in the chemical composition of their bones. Due to restrictions of time and available working equipment, only a brief consideration of this proposition was possible. The levels of three elements in the bones of the sheep were examined: calcium, magnesium, and zinc.

Osteological Microstructure

Background. The microstructural elements of bone consist of cells, organic matrix, and minerals. There are three types of bone cells: osteocytes, osteoblasts, and osteoclasts. These cells are primarily involved in laying down new bone and resorbing old bone (Vaughan 1970, pp. 23-60 and Comar and Bonner 1961, p. 610). In adult <u>Bos</u> bone, and presumably in <u>Ovis</u> bone as well, 88-89% of the organic matrix consists of collagen (Vaughan 1970, p. 61). The collagen is arranged in fibrils which are grouped into collagen bundles (Smith 1960). Bone mineral consists primarily of microcrystals of hydroxyapatite ($Ca_5(PO_4)_3OH$). The size and shape of hydroxyapatite crystals in bone has been a subject of considerable controversy.[2] The microcrystals are arranged in chains along the collagen fibres in an end to end relationship (Comar and Bonner 1964, p. 270) in such a way that the c-axes of the crystals are parallel to the long axes of the collagen fibres (ibid.; Caglioti, Ascenzi, and Santoro 1956, p. 426; Engstrom 1956, p. 7).

Crystalline Orientation. Drew, Perkins, and Daly (1971; MASCA 1970, 1973) have argued that one can distinguish wild from domestic caprines on the basis of crystalline orientation in bone. Examining thin sections of bone between crossed polarizers with a standard gypsum plate inserted in the light path and with the slow ray parallel to the c-axes of the apatite crystals, they maintain that thin sections of the weight-bearing bones of domestic animals showed strong blue interference colors on all articular surfaces. This pattern indicates that the crystals in these areas are almost all aligned with their c-axes perpendicular to the articular surface of the bone. The sections of bones from wild animals produced magenta interference colors in all areas of the bone, indicating that the hydroxyapatite crystals are basically randomly oriented throughout most of the bone (Drew et al. 1971, p. 281).[3]

Although this study was a provocative first step in the application of osteological microstructure to the differentiation of wild and domestic conditions, it had two problems. First, in the original study, they tested these techniques with archaeological samples which cannot be positively identified as being from either wild or domestic animals.[4] The material came primarily from two sites on the Anatolian plateau in Turkey: Suberde and Erbaba. On the basis of the analysis of the age-grade composition of the food animals' population, the character of the pattern of the remains indicating where the animals were killed, and the average size of individuals of each species, the caprines at Suberde and Erbaba were determined to be wild and domestic, respectively

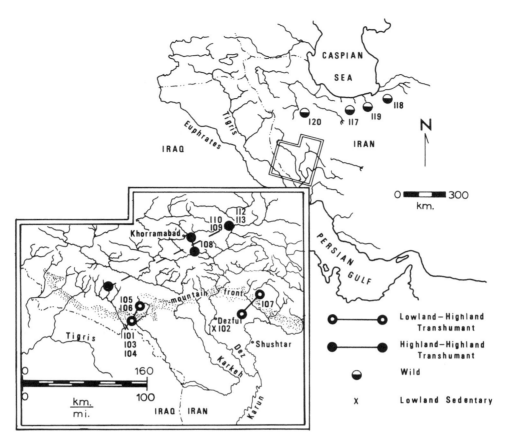

FIGURE 1. *Collecting localities of the lowland sedentary and wild sheep and the routes of the transhumant sheep used in the study.*

(ibid., pp. 280-281; and Perkins and Daly 1968). The use of age-grade compositions in the determination of wild and domestic animals is problematical at best (Clason 1972, p. 143). Moreover, reinterpretation of the age curve of the bones from Suberde by other researchers has led to the conclusion that they represent domestic herds (Wright and Miller 1976, p. 309). Finally, Uerpmann (in this volume) maintains that for reasons of time and site ecology it is "highly improbable" that the sheep of Suberde were wild.

The second problem in this study is that Drew et al. were not able to offer any concrete explanations for their results (1971, p. 282). They suggest (MASCA 1973, p. 1) "that the orientation of crystallites noted at the articular surfaces and in the shafts of the long bones developed as a response to stress in the weight-bearing bones of the bodies of domestic animals which, through lack of exercise, poor nutrition, or genetic deterioration, lacked sufficient material in their bones to form the sturdy bones characteristic of the wild animals studied." The claim, however, that early domesticates experienced such conditions is only conjecture and has not been substantiated. Moreover, the physiological mechanisms that, because of these conditions, would cause a strong preferred orientation of apatite in bone and the formation of thinner trabeculae (see below) of domestic animals remain to be demonstrated.

There are, however, theoretical grounds to expect differences in apatite orientation between animals with different levels and types of muscular activity. Living bone is an extremely plastic tissue, responsive to changes in mechanical load. Although the gross morphology of bone is genetically coded, it is mechanical loads which serve to refine and perfect the model and produce a functional skeleton (Murray 1936, pp. 19-20). Wolfe's Law of bone structure states that, "The form of the bone being given, the bone elements place or displace themselves in the direction of functional pressure and increase or decrease their mass to reflect functional pressure" (Bassett 1965, p. 18).

Not only will such modifications to mechanical load be reflected in the macrostructure of bone, but they should be even more apparent in the microstructural components. It has been argued that the orientation of collagen fibres and apatite crystals found on them is affected, if not determined by stress (Chatterji, Wall, and Jeffery 1972; Vaughan 1970, p. 20; Bassett 1965; Becker and Brown 1965, p. 1325; among others).

The mechanism behind the orientation of apatite and collagen in bone is not clear. Some researchers offer mechanical explanations. Studies show that, at birth, bone apatite and collagen fibres lack orientation (Winell, Bassett, and Spiro 1967); in other words that the formation of apatite and collagen is a random and continuing process without preferred orientation (Chatterji et al. 1972). With use, the bone is alternately stressed and relaxed. Under conditions of stress the apatite crystals unfavorably oriented with respect to the stress will dissolve into body fluid more easily than those more favorably oriented. During the relaxed state, both crystallization and crystal growth will occur, but more material will accrue to the favorably oriented crystals. This process will lead to differentiated crystal growth due to the greater thermodynamic stability of the larger crystals. The collagen fibres, on which the crystals are deposited, will also become oriented along lines of stress, whether due to packing considerations or to epitaxial formation of apatite. The proportion of favorably oriented crystals and collagen fibres will increase with use of bone, thereby making it more capable of withstanding stress. Chatterji et al. (1972, p. 157) conclude that as the degree of stress varies along the bone, so should the degree of orientation, and that bone subjected to higher degrees of stress should exhibit a higher degree of crystal orientation.

Other researchers point to the possible existence of piezoelectric effects in bone which, when excited by stress, cause changes in crystalline orientation. Some proponents of this theory believe that bone has semiconducting properties (Becker, Bassett, and Bachman 1964).[5] They feel that the collagen and hydroxyapatite interface may be a semiconductor of the PN type. A PN junction is one in which the relative availability of electrons is different in its components. Bending a PN junction creates an electric potential. There is an abundance of electrons in collagen and a lack of them in hydroxyapatite. Electricity is, thus, thought to be generated when there is stress at the collagen-hydroxyapatite interface (Basset 1965, p. 21; Becker and Brown 1965, p. 1325).

Studies _in vitro_ of collagen in a solution exposed to _electrical_ currents showed that collagen fibres form at right angles to the direction of the current (Bassett 1965, p. 22; Bassett, Pawluk, and Becker 1963). The orientation of the fibres is not changed when the current is turned off. Bands formed more rapidly and in a more orderly pattern under small, intermittent currents, such as might be produced with alternate stressing and relaxing of bone in normal movement.

It has been suggested that some stresses may excite osteoblastic activity and cause the building of bone, while others may excite the osteoclasts and cause resorption of bone (Frost 1964). Bassett and Becker (1962) and Bassett (1965, p. 23) maintain that the explanation behind the building and resorption of bone as a response to stress lies in its piezoelectric properties. Bassett (1965) demonstrates that regions under compression, which tend to be concave, are usually negatively charged. Such a region will be built up, with crystal orientation favorable to stress. A convex region, which is negatively charged, will be resorbed. He postulates that the orientation and formation of minerals and collagen in bone is a response to an electrical current and this current is determined by the strength and direction of applied force.

Whatever the explanation, most workers agree that orientation of apatite in bone is influenced by muscular stress.[6] Some researchers maintain that the sensitivity of apatite orientation may be so great as to reflect such small difference in muscular stress as a difference in gait (Bassett 1965, p. 25).

In light of the influence of mechanical load on crystalline orientation in the bones of the sixteen sheep examined here might be expected. Such differences would be the result of stresses of different types of muscular activity to which the bones were subjected as a consequence of the herd adaptations of the four groups of sheep. Therefore, rather than expecting differences between wild and domestic species as did Drew et al., variations in crystalline orientation due to contrasting herd adaptations and resulting muscular activity were predicted. It was predicted that the wild and the highland-highland transhumant sheep would have the most similar crystalline orientation because both inhabit mountainous areas year round. The distances traveled by the domestic highland sheep may cause some variation in orientation from the more localized wild sheep. The lowland sheep group, occupying relatively flat areas, was predicted to be the most different in crystalline orientation from the other groups, and the lowland-highland transhumants were expected to have crystalline orientation intermediate to the highland and lowland forms.

Standard petrographic thin sections of a distal humerus from each specimen were prepared.[7] Three sections were made from each bone: a longitudinal section cut parallel to the sagittal plane; a cross section cut perpendicular to the sagittal plane; and a section from the shaft cut parallel to the sagittal plane.

Examination of the transverse sections under polarized light with a standard quartz plate inserted in the light path revealed that, contrary to Drew et al.'s results, all specimens, wild and domestic, exhibited multiple bands on the articular surface, with crystals at right angles to the underlying band (Fig. 2).[8] This multiple banding is also apparent in pictures taken of this area with the use of a scanning electron microscope (Fig. 3).[9]

Drew et al. (1971, p. 281) maintained that the shafts of the domestic samples produced alternating bands of yellow and blue interference colors and that those from the bones of wild specimens appeared to be aligned: "in more or less concentric layers in which the apatite prisms are arranged radially as well as parallel to the long axis of the bone." I was unable to find any consistent differences in the crystalline orientation of the shafts of the wild and domestic specimens examined here, nor were there any differences between the different groups of domestic sheep.

The results obtained here agree with those of Watson (1975). According to Watson, thin sections of bones from wild and domestic modern sheep and goats all "showed a strong blue interference colour when the articular surface was oriented perpendicular to the slow direction of the gypsum plate." He also was not able to duplicate their results for the crystalline orientation of apatites in the shafts of his wild and domestic specimens.[10]

In summary, examination of longitudinal thin sections from the distal humeri of the wild sheep and the three groups of domestic sheep revealed some variability in orientation in different places on the same section and within groups of animals. However, it was not possible to detect any of the distinctive differences in crystalline orientation predicted between the groups examined. This is not to say that no such differences exist. Indeed there is some reason to expect that there will be variations in crystalline orientation of the bones of sheep with contrasting herd adaptations. It is likely that the negative results obtained here stem from the use of an inappropriate method for the measurement of these differences and/or the use of an inappropriate part of the skeleton for their detection. Examination of specimens using only polarized light, such as utilized here and by Drew et al., yields only vague, qualitative suggestions of crystalline orientation. Interference colors observed here might also be due to differences in the thickness of the sections as well as to differences in the density of bones. In order to determine if there are indeed significant differences in crystalline orientation in the bones of these animals, it is necessary to obtain quantitative data on the degree of orientation. Such measurements may be made with a universal stage attachment on a polarizing microscope. These measurements might also be taken using x-ray diffraction techniques.[11] Further studies with the scanning electron microscope might provide this information as well.

Trabecular Thickness. The second approach used here in the study of osteological microstructure involved the thickness of the trabeculae of the spongy bone of the distal humeri. Trabeculae are primarily found in the articular ends where bone is almost wholly loaded in compression. They are usually absent where bending stresses are appreciable. They function in the lateral transfer of compression loads from one bone to another, provide increased flexibility and damping of bone under load, and act as a means of transference of compression load from a large cross-sectional area (the joint) to a small cross-sectional area (the shaft) (Frost 1964, p. 38). The formation of trabeculae in these areas provides maximum longitudinal compressive strength and stability using the least amount of bone tissue (Arnold, Bartley, Tont, and Jenkins 1966, p. 34). It has been maintained that trabecular thickness is greatest where stress is greatest (Bourne 1956, p. 33).

Differences in muscular activity in animals with different herd adaptations should be expressed in varying compression forces on the bone and might, then, be reflected in trabecular thickness. Drew et al. (1971) maintain that the width of trabeculae in the articular ends of bone is greater in wild than in domestic caprines. However, for the reason discussed above, it may be better to look for these differences in terms of an animal's herd adaptation rather than in terms of its taxonomy. Due to probable high levels of compression stresses on bones of sheep which inhabit mountainous areas, the trabeculae of the wild and the highland-highland transhumants were expected to be thicker than those of the lowland sedentary sheep. The thickness of the trabeculae of the lowland-highland transhumants was predicted to be closest to that of the highland forms, especially to the highland-highland transhumants.

In order to test these predictions, photomicrographs were taken of the cross-sectional thin sections. These pictures were then enlarged 28 times (Fig. 4). Between five and eight areas showing minimal alteration due to sectional preparation were marked off on each cross section. The size of each area was measured with a polar planimeter. The intertrabecular spaces within the areas were also measured. To minimize measurement error, each measurement was taken five times and the results averaged. The sum of the areas of the intertrabecular spaces was then subtracted from the overall area measured to calculate the total amount of trabeculae in that part of the cross section. A ratio of the total amount of bone to space, the B/S ratio, was then calculated for each of the areas measured. The ratio is a measure of trabecular thickness. The thicker the trabeculae, the higher the ratio. Lower values of the B/S ratio indicate thinner trabeculae. Overall B/S ratios were then calculated for each bone and for each ecological group (Tables 1 and 2). Due to inferior quality of the section and/or the photomicrographs, one specimen from each group could not be accurately measured.

Statistical analysis revealed that the B/S ratios of the lowland-highland transhumants were significantly lower than those of the wild sheep (Table 3). They were also significantly lower than those of the highland-highland transhumants, although the significance was not as great. The ratios of the lowland-highland were lower than those of the lowland sedentary sheep, as well. However, this difference is not significant since there is a great deal of overlap between the upper ranges of the former and the lower ranges of the latter (Fig. 5). The difference between the lowland-highland transhumants and all other groups combined was significant. The B/S ratios of the lowland sedentary, the highland-highland transhumants, and the wild sheep could not be distinguished from each other.[12]

Since one of the functions of the trabeculae is the transference of compression load from a joint to a shaft, it was felt that trabecular width might be proportional to some index of the size and shape of the shaft. Unfortunately, the lower portion of the shaft, which might be more directly affected by compression stresses on the distal end of the bone, was not measured prior to sectioning. Measurements had to be taken on a more proximal portion of the shaft. Three measurements were taken, 2 to 3 cm below the deltoid tuberosity on the proximal shaft: a measurement of the anterior-posterior depth (L), a measurement of the lateral-medial breadth (W), and the thickness of the shaft on the anterior side directly

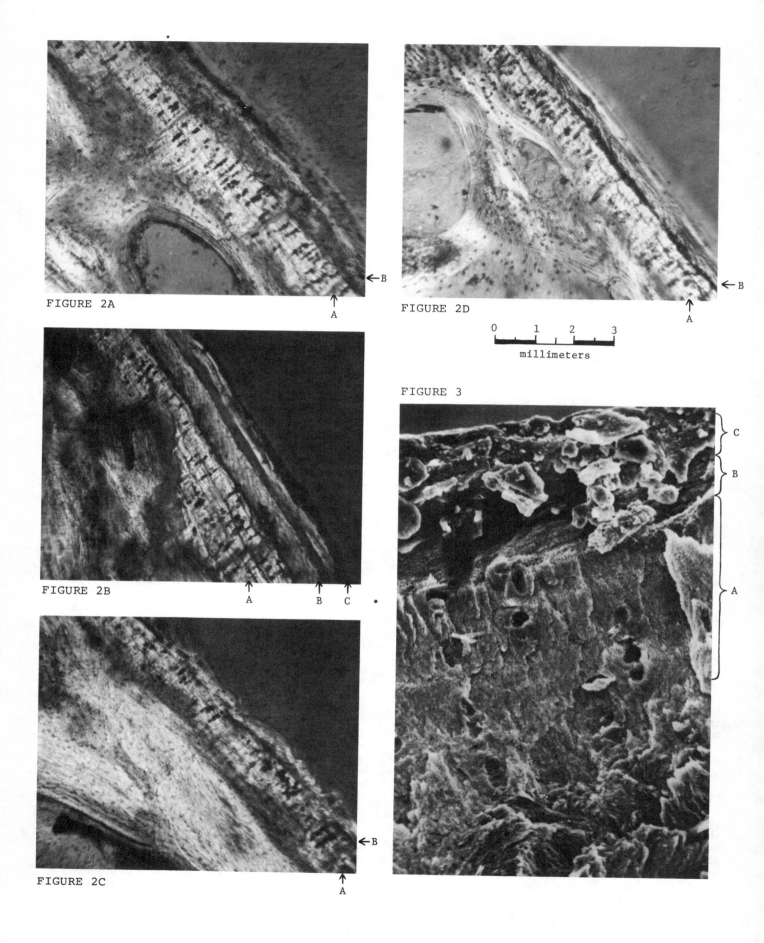

FIGURE 2A. *Longitudinal thin section of the distal humerus of specimen #103 (lowland sedentary domestic sheep) as viewed through a polarizing microscope. Multiple banding is evident. In each band, the crystals are oriented at right angles to the adjacent bands. The lower band (labelled 'A') with crystals perpendicular to the articular surface of the bone, corresponds to the "blue rim" reported by Drew et al. (1971 and MASCA 1970 and 1973).*

FIGURE 2B. *Longitudinal thin section of the distal humerus of specimen #105 (lowland-highland transhumant domestic sheep) as viewed through a polarizing microscope.*

FIGURE 2C. *Longitudinal thin section of the distal humerus of specimen #113 (highland-highland transhumant domestic sheep) as viewed through a polarizing microscope.*

FIGURE 2D. *Longitudinal thin section of the distal humerus of specimen #117 (wild sheep) as viewed through a polarizing microscope. The strong preferred orientation of crystals on the articular surface of the bone is clearly shown here.*

FIGURE 3. *Longitudinal cut from the distal humerus of specimen #112 (highland-highland transhumant domestic sheep) as viewed with a scanning electron microscope (SEM). Magnified 400 times. Portions of the bone lettered 'A', 'B', and 'C' correspond to the lettered portions of Figure 2 and denote the multiple banding at the articular surface of the bone. The outermost band 'C' is partially hidden by unevenly fractured bone.*

FIGURE 4. *Cross section from the distal humerus of specimen #108 (lowland-highland transhumant domestic sheep) viewed through a polarizing microscope. Magnified ca. 14.5 times. Thin sections such as these were used in the calculation of bone/space (B/S) ratios.*

below the deltoid tuberosity (T). The depth measurement was then multiplied by the breadth measurement to provide an index of the size of the shaft (LxW). The depth was also divided by the breadth to provide an index of the shape, or ovalness of the shaft (L/W) (Tables 1 and 2).

These measurements were grouped by herd type and compared using Student's t-test to see if any of these measurements could be used to discriminate between these groups. Results of these analyses revealed that only the LxW indices of the lowland sedentary sheep were significantly different from the other three groups. However, the ranges of the LxW measurements show overlap in the values of all three groups. Taking the B/S ratios as some function of the LxW, L/W, and the T measurements did not increase their ability to discriminate between groups.

To test the proposition that trabecular thickness is related to shaft size, shape, and thickness, correlation coefficients were computed between B/S and LxW, L/W, and T (Table 4). These tests revealed that the B/S ratios and the L/W index tend to be negatively correlated, especially in wild sheep. Such a relationship is expected if the trabeculae do indeed function in transference of compression load, maximizing tensile strength of bone while minimizing energy expenditure in bone manufacture, and if trabecular width is a function of the degree of stress.

TABLE 1. Mean and standard deviation of bone/space ratio (B/S) and shaft depth x shaft width (LxW in cm^2), shaft depth/shaft width (L/W), and thickness of shaft (T in cm) by specimen. [See Fig. 1 for habitat location of each specimen.]

Specimen Number	B/S	LxW	L/W	T
102	.84833 ± .22257	3.65	1.40	.415
103	1.27620 ± .36851	3.73	1.31	.325
104	1.30600 ± .51636	4.23	1.31	.325
105	.83167 ± .34493	2.20	1.87	.335
106	.96667 ± .42782	3.55	1.26	.225
108	.97600 ± .12542	2.20	1.41	.340
110	1.37200 ± .64033	2.85	1.41	.310
112	1.39330 ± .36615	2.94	1.17	.290
113	1.00400 ± .34602	3.44	1.26	.310
117	1.38330 ± .85099	2.56	1.22	.335
118	1.27000 ± .21423	2.91	1.36	.325
120	1.32820 ± .37934	2.80	1.26	.340

To briefly summarize the results of this aspect of the study, in the groups of animals studied here trabecular thickness can only be used to distinguish lowland-highland transhumants from wild sheep and perhaps between these animals and the combined B/S ratios of all the other groups. The B/S ratios of the lowland sedentary sheep are also significantly lower than those of the highland-highland transhumants, although this significance is not great. Trabecular thickness cannot be used to distinguish between wild and all domestic groups. None of the shaft measurements could be used to distinguish between the groups. However, there appears to be a general, although not significant, relationship between trabecular thickness and the degree of ovalness of the shaft.

TABLE 2

Mean and standard deviation of B/S
LxW, L/W, and T of the specimens by group*

Variable	Lowland Sedentary	Lowland-Highland	Highland-Highland	Wild
B/S	1.1435 ± .25607	.92478 ± .08077	1.2564 ± .21887	1.3272 ± .05666
LxW	3.7500 ± .33655	2.8575 ± .66319	3.1625 ± .31117	2.9025 ± .32623
L/W	1.3325 ± .07932	1.4425 ± .29568	1.2250 ± .04015	1.2725 ± .06759
T	.39750 ± .05637	.31500 ± .06096	.31350 ± .02241	.35625 ± .04626

*See text and Table 1 for definitions of abbreviations and units of measurement

TABLE 3

Significant Student's $\underline{t}$-tests (t) and corresponding ratios between sample variances (F) for two strata of bone/space (B/S) ratio

Variable		Strata: (1)	(2)	Test Statistic	DF*	Significance
Lowland-Highland(1) vs. Wild (2)	Mean Variance N	.92478 $.65329 \times 10^{-2}$ 3	1.3272 $.32101 \times 10^{-2}$ 3	t = -7.0647 F = 2.0323	4 2,2	.0021 .3298
Lowland-Highland(1) vs. Highland-Highland(2)	Mean Variance N	.92478 $.65329 \times 10^{-2}$ 3	1.2564 $.47905 \times 10^{-1}$ 3	t = -2.4622 F = 7.3431	4 2,2	.0695 .1199
Lowland-Highland (2) vs. All Other Groups(1)	Mean Variance N	1.2424 $.35609 \times 10^{-1}$ 9	.92478 $.65239 \times 10^{-2}$ 3	t = 2.7601 F = 5.4583	10 8,2	.0201 .1640

* DF = Degrees of Freedom (calculated separately for numerator and denominator in the F statistic)

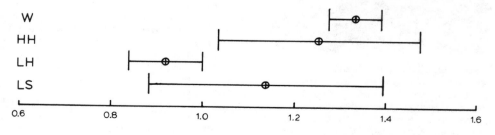

FIGURE 5. *Means and standard deviations of the bone/space ratios (B/S) for each ecological group: lowland sedentary (LS), lowland-highland transhumant (LH), highland-highland transhumant (HH), and wild (W).*

Caution must be used when considering these results. Examination of thin sections showed that trabecular thickness varies considerably over different cross-sectional areas of the articular ends of the humeri. In the future, care should be taken to assure that the location of sectioning is as exact and reproducible as possible. Furthermore, although an attempt was made to compensate for measurement error by taking each measurement many times, a more exact means of measurement should be found.[13] Moreover, due to the small sample from each ecological group considered here, any results reported can only be taken as suggestive, not definitive. It must also be remembered that trabecular thickness is purportedly related to compression and not to tension stresses (Frost 1964, p. 38). Differences in muscular activity of these animals may, however, be expressed in terms of tension and bending stresses, not in compression. If this is true, these differences might be reflected in areas of bone more likely to be affected by these stresses.[14] Finally, there are factors other than compression, age for example,[15] which affect trabecular thickness and must be considered.

TABLE 4. Correlation coefficients for B/S, LxW, L/W, and T of wild specimens*

Variable

B/S	1.0000			
LxW	-.9654	1.0000		
L/W	-.9820	.8988	1.0000	
T	.6934	-.4816	-.8171	1.0000
	B/S	LxW	L/W	T

* See text and Table 1 for definitions of abbreviations.

Chemical Composition. The last proposition considered was that differences in the sheep's diets would be reflected in the chemical composition of their bones. The main source of chemical elements in the tissues of grazing animals is dietary intake (Underwood 1962). Therefore, in order to evaluate the elemental make-up of these tissues, including bone, it is necessary to first examine the chemical make-up of the pasturage on which they graze.

The chemical composition of plants depends on several factors. First, the genus, species, or even strain of the plant are important factors (ibid., p. 369). Second, the nature of the soil in which the plant was grown is of great importance. A third factor is the climatic or seasonal conditions during growth. Finally, the stage of maturity of the plant influences its chemical content.

Differential intake of elements in pasturage is often reflected by differential representation of these elements in the various tissues of animals (ibid.). The intricate relationship between dietary intake of sulphates, copper, and molybdenum is well documented (ibid.; Maynard and Loosli 1956, p. 184; Marcilese, Ammerman, Valsecchi, and Dunavant 1969; Marcilese, Valsecchi, Rossi, Rudelli, and Figueiras 1970). High intakes of inorganic sulphates influence the level of copper and molybdenum retention in ruminants. Examination of the amounts of these elements in the bones of ruminants might, therefore, be used to key these animals to pastures where these conditions exist. It has also been suggested that the level of activity of sheep and the climatic conditions under which they live influence chemical composition of various tissues (Mukhtar 1970).

Interviews with herders of the domestic sheep collected for this study indicated that there are differences in the nutritional intake of the three groups of animals. The nutritional intake of the wild sheep, collected a considerable distance from the domestic animals (Fig. 1), should be quite different from the other groups analyzed. Studies of plant communities of lowland, steppe, and highland pastures (Pabot 1960) document considerable variation in the phytogeography of these areas. Therefore, in spite of the similarities of the highland topography of the ecozones of the wild and some of the domestic groups, it was predicted that the bones of the wild sheep should be the most different in elemental content because of probable differences in substrate and pasturage in the Elburz and Zagros upland pastures. While the substrate of the ecozones of the domestic sheep is fairly uniform, however, it was felt that differences in pasture plants might result in differential deposition of elements in the bones of the domestic groups. Specifically, the lowland sedentary and the highland-highland transhumants should be the most different, while the lowland-highland transhumants, whose ecozone overlaps with those of the other two groups, should have bones with chemical compositions intermediate between the lowland and highland forms.

The levels of calcium, magnesium, and zinc[16] in the bones of the Iranian sheep were measured using atomic absorption spectrophotometry (Tables 5 and 6). Statistical analysis of the percentage of these elements in the bones revealed that the levels of calcium in the bones of the lowland sedentary sheep were significantly higher than those of the lowland-highland transhumant sheep (Table 7). It was also indicated that the levels of calcium in the bones of the lowland-highland transhumants were significantly higher than those of the highland-highland transhumants. However, the significance of this difference is not as great. Moreover, there is some overlap between lowland-highland transhumants and the highland-highland sheep in calcium levels at one standard deviation from the mean (Fig. 6). There are no significant differences in the levels of magnesium or zinc among the groups of sheep. However, a Two-Way Student's t-test comparing the wild specimens with the domestic groups combined showed the former to be significantly higher in the level of magnesium in its bones than the domestic groups. Here again, there is some overlap at one standard deviation from the mean levels of magnesium in all groups (Fig. 3).

It must be stressed that these results must not be taken as any more than a demonstration that there may be significant differences in the chemical composition of the bones of caprines

TABLE 5

Mean and standard deviation
of Ca, Mg, and Zn by specimen

Specimen Number	Ca	Mg	Zn
101	59.67 ± 6.51	.49 ± .06	.02 ± .01
102	56.33 ±24.01	.56 ± .02	.06 ± .02
103	46.33 ± 9.02	.45 ± .15	.03 ± .01
104	50.67 ± 5.03	.41 ± .06	.03 ± .01
105	71.00 ± 9.54	.46 ± .01	.03 ± .01
106	67.67 ± 2.52	.49 ± .10	.02 ± .01
107	62.00 ± 7.94	.51 ± .09	.02 ± .00
108	73.00 ± 2.83	.52 ± .05	.03 ± .01
109	40.67 ± 2.31	.32 ± .03	.01 ± .001
110	54.33 ± 3.79	.43 ± .01	.04 ± .00
112	67.00 ±14.42	.54 ± .08	.03 ± .01
113	59.67 ± 3.79	.55 ± .07	.03 ± .00
117	61.67 ± 4.93	.58 ± .09	.02 ± .01
118	71.00 ±23.64	.65 ± .02	.03 ± .02
119	66.33 ±27.50	.61 ± .06	.03 ± .02
120	47.00 ±11.53	.45 ± .10	.01 ± .01

TABLE 6

Mean and standard deviation of Ca, Mg, and Zn by group

Element	Lowland Sedentary	Lowland-Highland	Highland-Highland	Wild
Ca	53.250 ±6.0759	68.500 ±4.7958	55.500 ±11.030	61.500 ±10.344
Mg	.47750± .06397	.49500± .02646	.45750± .10532	.57250± .08655
Zn	.03500± .01732	.02500± .00577	.02750± .01258	.02250± .00957

TABLE 7

Significant Student's t-tests (t) and corresponding ratios between sample variances (F) for two strata of mineral content (Ca and Mg)

Variable		Strata: (1)	(2)	Test Statistic	DF*	Significance
Ca Lowland-Highland(2) vs. Lowland Sedentary(1)	Mean Variance N	53.250 36.917 4	68.500 23.000 4	t=-3.9403 F= 1.6051	6 3,3	.0076 .3535
Ca Lowland-Highland(1) vs. Highland-Highland(2)	Mean Variance N	68.500 23.000 4	55.500 121.67 4	t= 2.1617 F= 5.2899	6 3,3	.0739 1.023
Mg Wild(2) vs. All Domestic(1)	Mean Variance N	.47667 .45879x10^{-2} 12	.57520 .74917x10^{-2} 4	t=-2.2966 F= 1.6329	14 3,11	.0374 .2383

* DF = Degrees of Freedom (calculated separately for numerator and denominator in the F statistic)

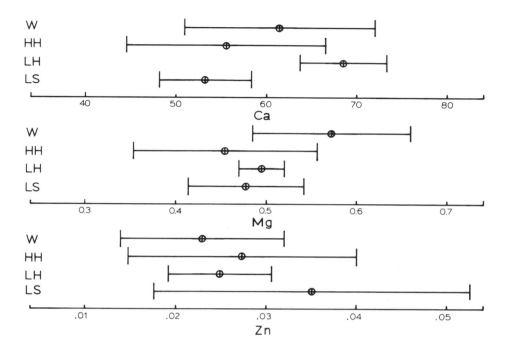

FIGURE 6. *Means and standard deviations of Calcium (Ca), Magnesium (Mg), and Zinc (Zn) content of the bones for each ecological group: lowland sedentary (LS), lowland-highland transhumant (LH), highland-highland transhumant (HH), and wild (W). Expressed as parts per hundred.*

from different environments. Due to restrictions of time and equipment, determination of the levels of elements such as copper, iron, sodium, potassium, selenium, molybdenum, and strontium, among others, was not possible. These elements may prove to be more informative than those measured here because they are not as critical in the physiological processes of bone formation and growth as calcium and magnesium. Therefore, these elements might better reflect the nutritional intakes and environments of the animals.[17]

Conclusion

Two propositions have been considered which may be used to guide the search for osteological differences between caprines which have participated in different ecosystems. The first was that the level and type of muscular activity of caprines with different herd adaptations will result in structural modifications in their bones. The consideration of this proposition was limited to the examination of two aspects of the microstructure of the bones of sixteen Iranian sheep from four ecological backgrounds. These aspects were crystalline orientation and trabecular thickness. In both cases certain predictions were made concerning the microstructural differences expected in the bones of these animals on the basis of information available on bone growth and function.

The results of the study of crystalline orientation of bone were contrary to those of Drew et al. (1971) which indicated marked differences in the weight-bearing bones of wild and domestic caprines. The predictions made here were also not supported, since no consistent differences in the crystalline orientation of the distal humeri of the wild and three groups of domestic sheep examined were found. These results do not mean that the examination of crystalline orientation will not eventually provide fruitful results in the establishment of differences between the bones of caprines with different herding adaptations, if appropriate techniques are used. The theoretical grounds on which to expect such differences have already been discussed.

There are three questions which should be considered before pursuing this line of research. First, what are the observable effects of different kinds of stress on apatite orientation in bone? Second, where and in which bones would differences in muscular activity as a result of contrasting herd adaptations be most strongly felt? Finally, are these differences sufficient to cause detectable and significant differences in crystalline orientation?

The second aspect of osteological microstructure considered in relation to the proposition that structural changes will result from differences in herd adaptation was the trabecular thickness in the articular ends of weight-bearing bones. Here again the results of this study contradict those of Drew et al. (1971). The predictions made here concerning trabecular thickness were only partially confirmed. The study conducted by Drew et al. (1971) indicated that the trabeculae of wild caprines are significantly thicker than those of domestic caprines. Although there was a general tendency for the trabeculae of the bones of the wild sheep to be thicker than those of the three domestic groups studied, as predicted, they were not significantly higher than those of the highland-highland transhumant sheep. However, an unpredicted result was that the trabeculae of the wild sheep were also not significantly higher than those of the lowland sedentary sheep, the group expected to have the thinnest trabeculae. Rather, the group with the thinnest trabeculae was the lowland-highland transhumants, predicted to be closer to the highland-highland forms in trabecular thickness.

None of the measurements of the shafts of the humeri, nor any of the functions of the Bone/Space ratios and the measurements could be used to discriminate between these groups. There was, however, a general, although not significant, tendency for the degree of ovalness of the shaft and the thickness of the trabeculae to be negatively correlated as might be expected if trabeculae truly function in the transference of compression load from the shaft to the joint. This tendency was more marked in the wild sheep whose bones were predicted to undergo more compression stresses than some of the other domestic groups.

Before we can properly generate and test hypotheses regarding the variation of trabecular thickness of the articular ends of caprine weight-bearing bones as a function of compression stresses on the bones, information on muscular activity of caprines with different herding adaptations and the different reaction of bone to stress must be gathered. Again three questions must be considered before proceeding with this research. First, what, if any, differences in compression stresses are there between these animals? Second, what, if any, is the relation between trabecular thickness and compression forces? Finally, what is the effect of other factors, such as age, on trabecular thickness and how can they be controlled.

The second proposition considered here was that differences in nutritional intake of caprines living in different environments will result in contrasting chemical compositions of their bones. Again certain predictions were made concerning the differences expected on the basis of substrates and the pasturage found in the ecozones which these animals inhabited. These predictions were not wholly supported with the elements tested here. The only truly positive result of this study was that the levels of calcium were significantly higher in the bones of the lowland-highland sheep than in those of the lowland sedentary sheep, groups expected to be fairly similar. There may also be some differences in the calcium levels of the bones of the lowland-highland transhumants and the highland-highland sheep, another unexpected result. In addition, the levels of magnesium in the bones of the wild sheep were significantly different from those of all the domestic animals when considered as one group. Such a result was predicted. It is believed that measurement of the levels of other elements in bone which are less involved in bone growth and maintenance will be more rewarding. It is also felt that other analytical techniques, such as neutron activation, would be preferable to atomic absorption in this study.

Before any direct links between nutritional intake and chemical composition of the bones of caprines may be drawn, more detailed information about the plant assemblages and nature of the soil in pastures of different environmental zones in the area, as well as geological and climatic data, should be compiled. If chemical composition is to be used to distinguish between ecologically different groups of animals, elements must be found which can be used to key an animal to a specific environment, upland pastures for example, yet which are not so localized as to be present in only a certain pasture in that area. Also, consideration must be made of the validity of the assumption that present-day diets of different groups of caprines are the same as those of ancient herds. For example, it is likely that the distribution of wild sheep in the past was much greater than that of modern wild sheep. As a result of hitherto unrestricted hunting and current conservation measures, wild sheep are restricted to high mountain pastures, nonrepresentative of the distribution of their ancestors. Age and differences in mineral metabolism at different stages of the animal's life cycle should also be considered. In addition, alteration of the chemical composition of archaeological bone may occur due to ion exchange with the surrounding substrate. If such factors can be controlled, study of the chemical composition of bone may prove to be an extremely useful means of distinguishing between animals from different ecosystems.

In conclusion, concrete distinctions between caprines from these different ecosystems could not be established. However, it must be stressed that the present study was only a beginning in the utilization of the techniques employed here in zooarchaeological research. What this study has accomplished is to draw an outline of the direction for future research. It is hoped that, as a result, a clearer definition of the types of information needed for such studies, the techniques available to conduct them, as well as their potential for success, has been gained.

Most importantly, it has been demonstrated that by approaching this problem as a question of the relation of an animal to its environment, it will be possible to generate, test, and refine hypotheses concerning osteological differences in animals resulting from this relationship. It is becoming increasingly apparent that we can no longer rely only upon traditional taxa-based identifications to provide answers to the questions which are being asked of zooarchaeology today. This is especially true of the problem of the origin of domestication, a topic which has been the focus of many of the efforts to establish taxa-based identification criteria. It is not sufficient to approach the differences between wild and domestic animals on the basis of specifics alone. It is likely that highland domestic sheep resemble their wild cousins in behavior and physiology more than they do their lowland domestic brothers and that, as a consequence, entirely different strategies are required in the herding of these "domestic" animals. What is meant by domestication in each case is not the same question at all.

This is not to belittle the accomplishments made in the establishment of taxa-based identifications.

The work in this area has been of immense value to zooarchaeological research. It has also provided the foundation upon which ecologically based identifications may be made and, thus, upon which new, more far-reaching problems may be approached. Having begun this line of research, it is now possible to formulate new and more refined ways to investigate the problem of the discrimination between the bones of caprines from contrasting ecosystems.

The success of this effort would have broad implications for the use of zooarchaeological research in the study of the evolution and operation of cultural systems in the Near East. For all periods of cultural development, the ability to discriminate between caprines from different ecosystems would allow application of zooarchaeological research to vital questions of social system evolution and operation. In the past, zooarchaeologists have not fully utilized the potential of faunal material in the study of culture change. Perhaps one reason for this lies in the lack of techniques which allow questions with broader cultural implications to be asked of the data. It is only through the development of the ability to extract from faunal material the sort of detailed information sought here that the far-reaching applications of zooarchaeology to anthropological research will be realized.

Notes
1. The sheep were purchased with funds awarded by the Honors Council of the University of Michigan and the Undergraduate Research Grant of the Anthropology Department of the University of Michigan.
2. For differing opinions see Caglioti, Ascenzi, and Santoro 1956; and Engstrom 1956.
3. These results differ from those first reported in MASCA 1970. At that time, Drew, Perkins, and Daly maintained that the bones of domestic animals show positive elongation of apatite crystals, while those of wild animals show negative elongation. Positive crystals produce blue interference colors when their c-axes are parallel to the slow ray of a gypsum plate. Negative crystals produce yellow interference colors when their c-axes are parallel to the slow ray (Rogers and Kerr 1942, p. 92; Wood 1964, p. 180). According to Drew et al. (MASCA 1970) the bones of the domestic animals produce yellow interference colors and those of wild specimens produce blue interference. Although they neglected to mention the orientation of the slow ray of the gypsum plate, I assume it was aligned perpendicular to the c-axes of the apatite crystals in the bones, given the birefringent qualities of positive and negative crystals.

They postulated (MASCA 1970) that this difference in "sign of elongation" may result from a difference in the habit of the apatite crystals of the bones of wild and domestic caprines. The crystals characteristic of bone may either assume a prismatic or tabular habit; prismatic crystals show negative elongation, while tabular crystals show positive elongation. Drew et al. (MASCA 1970) point to this fact, implying that the crystals of domestic animals are tabular in habit while those of wild animals are prismatic.

However, Drew et al. (MASCA 1970) appear to be confusing sign of elongation with optic sign. Although prismatic and tabular apatite crystals differ in sign of elongation they are both optically negative and have the same birefringence (Kraus, Hunt, and Ramsdell 1959, p. 352; Kerr 1959, p. 230; Deer, Howie, and Zussman 1966, p. 504). This confusion of optical terminology is also apparent in Drew et al.'s (1971) rebuttal to an article by McConnell and Foreman (1971) which questioned the results they obtained with x-ray diffraction studies. Here they maintained that in the bones of domestic animals the tabular crystals, oriented with their basal plains parallel to the articular surface of the bone, and thus with their c-axes perpendicular to the surface, produce strong blue interference colors when the slow ray of the gypsum plate is perpendicular to the articular surface; i.e., parallel to the c-axes. If the birefringent properties of the bone are produced by the crystals alone, such a result could only be produced if the crystals were optically positive, which they are not.

The only way that the results reported in the first MASCA publication (1970) could be obtained is if the c-axes of the crystals in the domestic and wild specimens were oriented in opposite directions to the articular surface of the bones. Drew et al. (MASCA 1970) do offer differences in the orientation of the "long axes" of the apatite crystals as an alternate explanation for the results reported here. It is unclear if these "long axes" refer to the c-axes or the axes of elongation. However, neither explanation is mentioned in subsequent publications where it is maintained that the apatite crystals have strong preferred orientation in domestic animals and are randomly oriented in wild animals.
4. In later publications Drew (Pollard and Drew 1975 and MASCA 1973) says that these techniques have been extended to modern samples with positive results. However, no clear pictures of these bones are available. It is also interesting to note that in her latest publication on this research the random orientation of the wild specimens is no longer mentioned and it is the greater degree of orientation in the domestic specimens that is the primary distinguishing characteristic between wild and domestic animals.
5. For other theories about piezoelectric properties of bone, see Shamos and Lavine 1964 and 1967; Fukada and Yasuda 1957; and Gjelsvik 1973.
6. Other researchers in the field do not agree with this position. Some feel that the type of stress bone is subjected to will not induce structural changes (Richard Taylor, personal communication), and feel that the increasing degree of apatite and collagen orientation after birth is a growth phenomenon, unrelated to extrinsic factors of stress.
7. The decision to prepare sections from the humeri and to observe the crystalline orientation of the distal articular surface and the shaft of these bones was made in an effort to duplicate the study of Drew et al. as closely as possible, using the modern specimens whose life history was more certain, in an effort to provide results most comparable to their work with archaeological bones.
8. In all the sections made for this study, the multiple banding was evident, varying to a certain

degree with the quality of the section. For the purposes of comparison, the innermost major band (blue) represents the area of the surface comparable to that discussed by Drew et al. The function of these multiple bands remains a mystery. Although some researchers who were shown the sections have suggested that they are an artifact of the sectioning technique (L. Katz, personal communication), this does not explain why they also appear in the SEM pictures. Other sections of modern pig bone and archaeological sheep and goat from the same area on the bone shown to me by R.H. Meadow do not contain them. It has been suggested that I use either the SEM or electron microprobe to determine the chemical composition of the bands (S. Garn, personal communication). It is also possible that these results are due to differences in density of the bone in this area.

9. Unfortunately, due to problems in the preparation of the samples for work with the SEM and lack of time, I was unable to view the wild bones using this technique.

10. Watson offers an alternative explanation for the results offered by Drew et al. that greatly aids in clarifying the problems with their study discussed in footnote 4. He maintains that the differences Drew et al. noted between wild and domestic archaeological specimens are the result of differences in collagen preservation after deposition of bone. He notes that quasi-crystalline collagen is optically positive while apatite crystals are negative. Thus, although the optic axes of apatite in bone are aligned parallel to the fibrils of collagen and to their optic axes, the negative birefringent qualities of the apatite is masked by the positive birefringence of the collagen. He concludes that the blue interference colors, obtained by Drew et al. when the slow ray to the gypsum plate is oriented parallel to the c-axes of the apatite crystals and thus to those of the collagen, is produced by the collagen and not the apatite. He supports this supposition by the fact that after the removal of bone mineral by decalcification, there was no change in the birefringence of the bone. The "wild" pattern of randomly oriented crystals might have resulted with decay of collagen after deposition and subsequent disorientation of the mineral crystals. This disorientation would result in birefringences that cancel out and produce magenta interference colors when viewed under crossed polarizers with a standard gypsum plate inserted in the light path.

11. Drew et al. (1971, p. 281) used x-ray diffraction techniques to confirm their results. There is some question about the validity of these results (McConnell and Foreman 1971). I did not use this technique since I was informed that it is an extremely difficult and time-consuming technique. However, I have been informed subsequently that this is not the case and that such techniques would provide quantifiable results on the degree of orientation of the apatite crystals in these bones (L. Katz, personal communication).

12. Drew et al. (1971, p. 281) also maintained that the shape of the intertrabecular spaces, which they refer to as "lacunae," of wild and domestic forms varied, with domestic animals having more rectangular spaces and wild, more rounded. Although they were not measured for shape, I found that the size and shape of the intertrabecular spaces varied greatly at different cross-sectional cuts in the bone and at different places within the same cut. I could detect no consistent differences in the size or shape of intertrabecular space between the groups of sheep I examined.

13. I have been told that such measurements might be taken by printing these pictures on special paper and then measuring light reflection from the paper (S. Garn, personal communication).

14. It must be reiterated here that not all researchers feel that bone structure is modified by external stress. This question has been and continues to be a major bone of contention.

15. It has been demonstrated that trabecular thickness may be used to age human bones (Kerley 1965). Age was not felt to be a significant factor here since most specimens were from animals 2 to 3 years of age.

16. An attempt was made to measure the level of copper in the bones. However, there was not enough copper in them to be detected which means a level of less than .003%. By increasing the amount of bone examined, copper could probably be measured using this technique.

17. It may, in fact, only be possible to demonstrate polymodality in the levels of trace elements in the bones of ancient herds, without being able to define the environmental type represented in the caprine population at any one site. However, such information, coupled with information yielded by other techniques, such as those considered here, might be of immense help in the definition of types of caprines represented.

Acknowledgments

Many individuals played important roles in various aspects of this study; without their help the research would not have been possible. For aid in collecting domestic specimens I would like to thank Rostam, Pourmand, Gorg-Ali Bakhtiari, Golam Rejaki, and above all Frank Hole and Sekander Amanolahi. I would also like to thank Luis delaTorre and the Field Museum of Natural History for loan of the wild specimens. Among the many individuals who gave invaluable assistance in the analysis of the bones of the animals, special thanks are due F. Gaynor Evans, Dominic Dziewaitkowski, James Hinchcliff, Martha Goodway, Peggie Hollingsworth, Myron Brownie, Betty Musgrave, Kathy Deffenbaugh, Marty Brown, and Eric Esscene. Helpful comments and criticisms on earlier drafts of this work were kindly provided by Christopher Peebles, C. Loring Brace, Richard Ford, and Richard Meadow. Above all I would like to thank Henry Wright and Richard Redding who have provided help and support in all aspects of this study. Finally, special gratitude is due to my husband M. James Blackman whose scientific and editorial criticisms have been essential in the distillation of the paper into its present form.

References

Arnold, J.S., M.H. Bartley, S.A. Tont, and D.P. Jenkins
   1966   "Skeletal changes in aging and disease," *Clinical Orthopedics*, vol. 49, pp. 17-38.

Bassett, C.A.L.
   1965   "Electrical effects in bone," *Scientific American*, vol. 213, no. 4, pp. 18-25.

Bassett, C.A.L. and R.O. Becker
   1962   "Generation of electrical potentials by bone in response to mechanical stress," *Science*, vol. 137, p. 1063.

Bassett, C.A.L., R.J. Pawluk, and R.O. Becker
   1963   "Effects of electrical currents on bone *in vivo*," *Nature*, vol. 199, pp. 1304-1305.

Becker, R.O. and F.M. Brown
   1965   "Photoelectric effects in human bone," *Nature*, vol. 206, pp. 1325-1328.

Becker, R.O., C.A.L. Bassett, and C.H. Bachman
   1964   "Bioelectrical factors controlling bone structure," in H.J. Frost, editor, *Bone Biodynamics*, pp. 209-232. Boston.

Bökönyi, S.
   1973   "Some problems of animal domestication in the Middle East," in J. Matolcsi, editor, *Domestikationsforschung und Geschichte der Haustiere*, pp. 69-75. Budapest.

Bökönyi, S., L. Kakkai, and J. Matolcsi
   1965   "Vergleichende Untersuchungen am Metacarpus des Urs- und des Hausrindes," *Zeitschrift für Tierzüchtung und Züchtungbiologie*, vol. 81, no. 4, pp. 330-347.

Boessneck, J., H.-H. Müller, and M. Teichert
   1964   *Osteologische Unterscheidungsmerkmale zwischen Shaf (Ovis aries Linné) und Ziege (Capra hircus Linné)*. Kühn-Archiv, vol. 78, no. 1-2, pp. 1-129.

Bourne, G.H.
   1956   *The Biochemistry and Physiology of Bone*. New York.

Caglioti, V., A. Ascenzi, and A. Santoro
   1956   "On the interpretation of the low-angle scatter of x-rays from bone-tissues," *Acta, Biochemica et Biophysica*, vol. 21, no. 3, pp. 425-432.

Chatterji, S., J.C. Wall, and J.W. Jeffery
   1972   "Changes in the degree of bone materials with age in the human femur," *Experimentia*, vol. 28, p. 157.

Clason, A.T.
   1972   "Some remarks on the use and presentation of archaeozoological data," *Helinium*, vol. 11, no. 2, pp. 140-153.

Comar, C.L. and F. Bonner
   1961   *Mineral Metabolism*, vol. 1, part B. New York.
   1964   *Mineral Metabolism*, vol. 2, part A. New York.

Deer, W.A., R.A. Howie, and J. Zussman
   1966   *An Introduction to the Rock-Forming Minerals*, pp. 504-509. New York.

Degerbøl, M.
   1963   "Prehistoric cattle in Denmark and adjacent areas," in A.E. Mourant and F.E. Zeuner, editors, *Man and Cattle*. Occasional Paper No. 18 of the Royal Anthropological Institute, pp. 68-79.

Drew, I., D. Perkins, and P. Daly
   1971   "Prehistoric domestication of animals: effects on bone structure," *Science*, vol. 171, no. 3968, pp. 280-282.

Engstrom, A.
   1956   "Structure of bone from anatomical to the molecular level," in G. Wolstenholme and C.M. O'Conner, editors, *Bone Structure and Metabolism*, pp. 3-10. Boston.

Flannery, K.V.
   1961   "Skeletal and Radiocarbon Evidence from the Start and Spread of Pig Domestication." Master's Thesis, University of Chicago, Department of Anthropology.
   1969   "The animal bones," in F. Hole, K.V. Flannery, and J. Neely, *Prehistory and Human Ecology of the Deh Luran Plain*. Memoirs of the Museum of Anthropology, University of Michigan, no. 1, pp. 262-330. Ann Arbor.

Frost, H.M.
   1964   *The Laws of Bone Structure*. Springfield, Illinois.

Fukada, E. and I. Yasuda
   1957   "On the piezoelectric effect in bone," *Journal of Physics Society of Japan*, vol. 12, p. 1158.

Gjelsvik, A.
   1973   "Bone remodeling and piezoelectricity-II," *Journal of Biomechanics*, vol. 6, pp. 187-193.

Gromova, V.
   1953   *Osteologiceskie Otlicigja Rodov Capra (kozly) i Ovis (barany)*. Trudy Komissii po izučeniju cetvierticnogo perioda, vol. 10, no. 1. Moskva-Leningrad

Jewell, P.
   1963   "Cattle from British Archaeological Sites," in A.E. Mourant and F.E. Zeuner, editors, *Man and Cattle*. Occasional Paper No. 18 of the Royal Anthropological Institute, pp. 80-101.

Kerr, P.F.
   1959   *Optical Mineralogy*, pp. 230-231. New York.

Kerley, E.R.
   1965   "The microscopic determination of age in human bone," *American Journal of Physical Anthropology*, vol. 23, no. 2, p. 149.

Kraus, E.H., W.F. Hunt, and L.S. Ramsdell
   1959   *Mineralogy: An Introduction to the Study of Minerals and Crystals*, pp. 351-353. New York.

Marcilese, N.A., L.B. Ammerman, R.M. Valsecchi, and B.G. Dunavant
   1969   "Effect of dietary molybdenum and sulphate upon copper metabolism in sheep," *Journal of Animal Nutrition*, vol. 99, p. 177.

Marcilese, N.A., R.M. Valsecchi, F.M. Rossi, M.D. Rudelli, and H.D. Figueiras
   1970   "Review of research on mineral metabolism and diseases in farm animals in Argentina," in *Mineral Studies with Isotopes in Domestic Animals,* pp. 67-80. Vienna.

MASCA (Museum Applied Science Center for Archaeology, University of Pennsylvania)
   1970   "Bone from domestic and wild animals: crystallographic differences," *MASCA Newsletter,* vol. 6, no. 1, p. 2.
   1973   "Technique for determining animal domestication based on a study of thin sections of bone under polarized light," *MASCA Newsletter,* vol. 9, no. 2, pp. 1-2.

Maynard, L.A. and J.K. Loosli
   1956   *Animal Nutrition,* 4th edition. New York.

McConnell, D. and D. Foreman
   1971   "Texture and composition of bone," *Science,* vol. 172, p. 972.

Mukhtar, A.M.S.
   1970   "Some aspects of mineral nutrition problems in domestic livestock in the Sudan," in *Mineral Studies with Isotopes in Domestic Animals.* pp. 183-188. Vienna.

Murray, P.D.F.
   1936   *Bones.* New York.

Pabot, H.
   1960   The Native Vegetation and its Ecology in the Khuzistan River Basins. Mimeographed M.S.

Perkins, D. and P. Daly
   1968   "A hunters' village in neolithic Turkey," *Scientific American,* vol. 219, no. 5, pp. 97-105.

Pollard, G.C. and I. Drew
   1975   "Llama herding and settlement in prehispanic northern Chile," *American Antiquity,* vol. 40, no. 3, pp. 296-305.

Reed, C.
   1960   "A review of the archaeological evidence on animal domestication in the prehistoric Near East," in R.J. Braidwood and B. Howe, *Prehistoric Investigations in Iraqi Kurdistan.* Studies in Ancient Oriental Civilizations, no. 31, pp. 119-145. Chicago.

Rogers, A.F. and P.F. Kerr
   1942   *Optical Mineralogy.* New York.

Shamos, M.H. and L.S. Lavine
   1967   "Piezoelectricity as a fundamental property of biological tissues," *Nature,* vol. 213, pp. 267-269.

Shamos, M.H., L.S. Lavine, and M.I. Shamos
   1964   "Physical bases for bioelectric effects in mineralized tissues," *Clinical Orthopedics,* vol. 35, pp. 177-188.

Smith, J.W.
   1960   "Collagen fibre patterns in mammalian bone," *Journal of Anatomy,* vol. 94, no. 3, pp. 329-344.

Underwood, E.J.
   1962   *Trace Elements in Human and Animal Nutrition,* 2nd edition. New York.

Vaughan, J.M.
   1970   *The Physiology of Bone.* Oxford.

Watson, J.P.N.
   1975   "Domestication and bone structure in sheep and goats," *Journal of Archaeological Science,* vol. 2, pp. 375-383.

Winell, C.A., C.A.L. Bassett, J. Wiener, and D. Spiro
   1967   "Ultrastructural aspects of fibrogenesis and osteogenesis in tissue cultures," *Anatomical Record,* vol. 158, pp. 75-88.

Wood, E.
   1964   *Crystals and Light: An Introduction to Optical Crystallography.* Princeton.

Wright, G.A. and S. Miller
   1976   "Prehistoric hunting of new world wild sheep: implications for the study of sheep domestication," in C.E. Cleland, editor, *Cultural Change and Continuity,* pp. 293-312. New York.

PART FOUR

PAPERS ON FAUNAL REMAINS FROM SHAHR-I SOKHTA

Because the contents of the Shahr-i Sokhta papers cover many different themes, the editors were of two minds about how best to include these contributions in the volume. There was much discussion about having the authors slightly modify their papers so that portions of the preliminary remarks would fit into the part on methodology, the essay on the onager serve as an example of the use of measurements, and the camel paper stand as part of a separate section on domestication which might also have included the paper by Uerpmann now to be found in Part Two. We decided, however, that such an approach created more problems than it solved and that it would be best to establish a separate part of the volume for these papers, none of which really can stand alone without the "Preliminary Remarks."

In our introductions to Parts One, Two, and Three, we have referred to one or another of the Shahr-i Sokhta papers. Here we should note that from a zooarchaeological point of view, the morphological descriptions and osteometric data presented for each taxon represent a first step toward characterizing the ancient fauna native to eastern Iran, an area even less well-known than the admittedly poorly studied regions farther west. In particular, the papers on the camel, onager, and gazelle show the difficulties involved in trying to distinguish between closely related species belonging to the same genera. This problem is compounded by the fragmentary nature of the small archaeological sample involved and the limited number of specimens available for comparison, many of which, both prehistoric and modern, come from areas far from eastern Iran. As Boessneck and von den Driesch noted in their paper in Part Two, only after a long period of accumulating basic osteological data from many regions will it be possible to make more securely documented statements about the nature of and changes in the ancient animal populations of the Middle East.

Of the papers in this section, the contribution of Compagnoni and Tosi on the camel is of particular interest because it deals with a secondary domesticate which provided the key to man's use of deserts as highways for commerce. The preservation, recovery, and identification of camel hair and camel dung at Shahr-i Sokhta provide a unique insight into the early exploitation of these animals. By attempting to outline the distribution of the two camel species primarily using evidence from figural representations, the authors pose interesting questions which can only be met by future archaeological, zoological, and paleontological work throughout the Middle East.

R.H.M.
M.A.Z.

PRELIMINARY REMARKS ON THE FAUNAL REMAINS FROM SHAHR-I SOKHTA

Lucia Caloi, Istituto di Geologia e Paleontologia
University of Rome
Bruno Compagnoni, Servizio Geologico d'Italia
Rome
Maurizio Tosi, Istituto Universitario Orientale
Naples

Acknowledgments

We wish to acknowledge the kind help provided us in various stages of our work by Prof. A. Malatesta and the Istituto Italiano di Paleontologia Umana, the Department of Environmental Conservation of Iran, The British Museum (Natural History), the Museum of Comparative Zoology of Harvard University, the Field Museum of Natural History at Chicago, and the Museo Civico di Zoologia at Rome.

Particular gratitude is due the President of the Istituto Italiano per il Medio ed Estremo Oriente, Prof. Giuseppe Tucci.

Thanks to G. Silvestrini of IsMEO for taking the photographs and to Miss I. Reindell for drawing the figures.

Introduction

The first five campaigns of excavation carried out at Shahr-i Sokhta (Sistan) from 1967 to 1972 by the Italian Archaeological Mission in Iran led to the collection of an enormous quantity of bone remains from mammals, birds, reptiles, amphibians, and fishes as well as large numbers of eggshell fragments, arthropod exoskeletons, and marine and freshwater shells (Biscione et al. 1974, pp. 50-52).

In general, the bones collected were in a fairly good state of preservation, particularly the smaller examples, although, in some cases, the high salt content of the soil made the material very brittle owing to saltpeter crystals having penetrated the bone tissue. It was thus necessary to subject the specimens to repeated washings in running water before consolidating them with polymer resins.

Following a preliminary selection of the excavated material, about 20,000 bones from macromammals have, so far, been determined as coming from the following species:

Canis familiaris L. (domestic dog)
Vulpes vulpes L. (red fox)
Lutra lutra L. (otter)
Felis cf. catus L. (cat)
Equus hemionus Pallas (Persian half-ass)
Camelus sp. (camel)
Gazella subgutturosa Guldenstaedt (goitered gazelle)
Capra hircus hircus L. (domestic goat)
Capra hircus cf. aegagrus L. (wild goat)
Ovis aries L. (domestic sheep)
Ovis vignei Blyth (urial sheep)
Bos indicus L. (zebu cattle)

For the purposes of the present work, only the relatively less well-represented genera (excluding dog) have been examined, i.e., Vulpes, Lutra, Felis, Equus, Camelus, and Gazella, which, when taken together, account for less than one percent of the macromammal bones so far recovered from Shahr-i Sokhta. The priority given to the presentation of the less abundant fauna is justified by two considerations. First, even a preliminary survey of the genera with high socioeconomic importance in an urban or proto-urban complex (e.g., Capra, Ovis, and Bos) requires a proportionately greater effort. Although the results of such studies will be published in later papers, we think that it is worth mentioning here that, as a first approximation, the genera Capra, Ovis, and Bos account for 23.5%, 54%, and 21.5% of the total number of bones, respectively. Second, our decision to describe the less abundant fauna here is due to the important role which the remains of these species play in reconstructing the environmental conditions surrounding the protohistoric settlement of Shahr-i Sokhta. Because significant shifts in the course of the Hilmand delta have occurred over the past five thousand years, major changes in the distribution of human, animal, and plant populations have taken place in Sistan. Thus it is necessary to turn to the faunal and other paleobiological data to help to determine the nature of the ancient environment in the area of the site.

Apart from considerations of the environment, individual attention to certain of the less well-represented species is necessary because of the special zoological and cultural problems associated with them. In particular, the problem of the domestication of species of the genera Equus and Camelus in the fourth and third millennia B.C. of the Middle East must be discussed in light of the remains from Shahr-i-Sokhta. Both of these ungulate mammals are employed primarily in the transportation of people and goods and are not connected with food production or capitalization of surpluses as are the genera Ovis, Capra, and Bos which had long been domesticated by this late date. Equus and Camelus are linked to the transportation of goods, to the socioeconomic control over the open spaces of the Eurasian steppes and the subtropical deserts, and to the consequent expansion of economic interchange.

The Site

Shahr-i Sokhta covers an area of about 150 hectares on a terrace of Neogene alluvial deposits related to the zone of maximum expansion of a landlocked lake basin (Jux and Kempf, in press). The site is sharply divided into areas which saw varying degrees of human habitation or use. Areas of greater use are clearly visible because of their greater elevation and the clayey nature of the soil due to the accumulation of mud bricks from ancient buildings (Biscione et al. 1974, pl. 15a). The location was first inhabited about 3200 B.C. From its original area of about 15 hectares, the occupied portion of the city expanded until it reached a size of 80 hectares during its period of greatest importance (2400-2100 B.C.) at which time it was the center for intense demand and for economic interchange with the Afghan plateau, the Persian Gulf, Central Asia, and probably southern Iran (Tosi and Piperno 1973). A further 21 hectares were occupied by a graveyard which developed on the southern side of the terrace (Piperno and Tosi 1975, p. 187). At the end of the third millennium B.C. the settlement gradually declined as the result of a process of socioeconomic decentralization manifested mainly in demographic segmentation leading to a reduction in the urban area occupied and to a parallel increase in the number and density of farming villages. By the beginning of the second millennium, all that was left of Shahr-i Sokhta were a few dwellings which in turn were abandoned about 1800 B.C.

The cultural sequence of Shahr-i Sokhta is based on comparison of the stratigraphic seriations established in the various trenches dug within the inhabited areas, with the associations being checked with the grave furnishings. Combining the results of absolute dating techniques (radiocarbon, palaeomagnetism, Uranium 238 fission tracks) and of comparisons of the material culture with that from sites in the surrounding regions has permitted a chronological-stratigraphic reconstruction to be made (Table 1). A general chart showing the chronological distribution of the less well-represented macromammal finds has also been developed (Table 2).

Specimen numbers

The specimen numbers indicate provenience of the finds. They usually contain a determinant of horizontal provenience (*geodetic grid sector*, consisting of three capital letters of the Latin alphabet, or a reference to the *room* in which found, marked by a progressive Roman numeral) and of vertical provenience (10-15 cm artificial *cuts* indicated by means of progressive numerals from top to bottom and depositional *strata*, again starting from the top, represented by circumscribed numerals). The use of the abbreviation "s.c." (=uncertain) in the case of a number of bones is due to a mix-up of part of the bone material and related cards during transport from Iran to Italy.

Comments on recovery procedures

What seems to distinguish Shahr-i Sokhta from other contemporary sites in the Middle East is an exceptional degree of preservation of most perishable materials. Finds of skin, hair, dung, fibers, eggshells, and chitina (insects), far from being rare, occur in every sector of the site--both settlement and graves (Biscione et al. 1974, pp. 50-52). Previous papers have stressed how important it is to systematically collect such finds and have suggested that such remains could be recovered even in sites demonstrating lesser degrees of preservation than Shahr-i Sokhta (Costantini and Tosi 1975, pp. 312-313). Here we would like to proceed further and emphasize the value of such finds for zooarchaeological studies.

The recovery of specimens from earlier excavations was limited strictly to the bones of large mammals. Although we do not underestimate the importance of skeletal remains, we do believe that such data can be enriched almost everywhere by other kinds of remains such as animal hair, dung, skin, and eggshells as well as by other classes of fauna such as arthopods, mollusks, reptiles, and amphibians. While the skeletal parts of large mammals may be difficult to quantify because of statistical unreliability brought on by differential and accidental states of preservation, the very recovery of the other types of remains depends upon the conscious intent of the excavator. Time and money-consuming recovery techniques have to be systematically applied to retrieve microscopic finds, and once recovered, they have to be appropriately separated and identified. The last requirement--that of identification--is the most problematical because specialists are rare, often overwhelmed by work, and unable because of a lack of appropriate administrative structures to work full-time alongside the archaeologist. Interaction is therefore minimal just where it should be maximized to throw new light on the interaction between human culture and the environment.

Even given these problems, how can we begin to investigate the *Tierwelt* of a specific environment, bound to human exploitation and demographic concentration, when we exclude, *a priori*, entire groups of animals which are, after all, the ones best suited to identify those changes over space and time which are induced by the human economy? Notwithstanding the size of the problems involved, the archaeologist should make a point to recover these microscopic remains and should himself seek appropriate training to be able to proceed with preliminary separation of the finds in order to distribute them among the different specialists. Field archaeologists and, when available, field

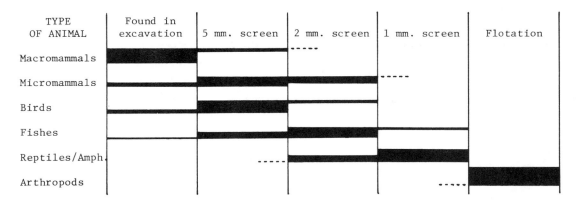

FIGURE 1. *Approximate proportions of faunal remains recovered using different techniques.*

TABLE 1: Shahr-i Sokhta, the chronological-stratigraphic sequence

Cultural Period	Structural Phase	Dating (B.C.)	Area of town (approx.)
IV	0	ca. 1800	?
	1	1900–1800	5 ha.
	2	2100–1900	?
III	3	2300–2100	80 ha.
	4	2400–2300	80 ha.
II–III	5	2500–2400	75 ha.
II	6	2600–2500	75 ha.
	7	2700–2600	?45 ha.
I	8	2900–2700	15 ha.
	9	3000–2900	15 ha.
	10	3200–3000	15 ha.

TABLE 2: Shahr-i Sokhta, chronological distribution of the less well-represented macromammal finds

Period:	IV/			III/			/	II/			I		
Phase:	0	1	2	3	4	5	6	7	8	9	10	?	T
Camelus	–	–	–	–	–	–	4	1	–	–	–	–	5
Equus	1	–	–	–	–	1	1	–	–	10	1	7	21
Gazella	–	2	–	–	1	7	8	1	2	1	–	16	38
Canis	–	–	–	–	–	4	4	2	1	4	–	7	22
Lutra	–	–	–	–	–	–	–	1	–	–	–	1	2
Felis	–	–	–	–	–	1	1	–	–	–	–	–	2
Vulpes	–	–	–	–	–	–	1	–	–	–	–	–	1
Totals	1	2	–	–	1	13	19	5	3	15	1	31	91

ethnobotanists must be the ones to collect paleobiological remains.

The recovery of paleobiological remains is extremely simple when compared with problems of identification and interpretation. Recent proposals by British scholars to increase the sophistication of recovery/separation devices in order to permit mechanical separation of all biological finds has been criticized elsewhere (Costantini and Tosi 1975, pp. 312–313). Experiments conducted at several excavations have permitted an association to be made between screen size and kind of fauna recovered. This is illustrated in Figure 1. Such a diagram helps the archaeologist visualize what limitations he is placing on his ability to reconstruct the animal world as preserved in settlement deposits by not using different recovery procedures. Further and more detailed discussions could fill out specific details by documenting the recoverability, given certain techniques, of each bone type or animal species, but the principle remains the same. What is emphasized here is that the archaeologist must take it upon himself to ask the proper questions beforehand, to plan and execute the recovery of paleobiological remains, and to be aware of and note features encountered during excavation which might affect the quantity and quality of the samples recovered. It is the archaeologist who is primarily responsible for the understanding of structural and distributional patterns and thus multidisciplinary approaches will not decrease but will increase the involvement of the archaeologist in the documentation and interpretation of the past.

References

Biscione, R., G.M. Bulgarelli, L. Costantini, M. Piperno, M. Tosi
  1974 "Archaeological discoveries and methodological problems in the excavations of Shahr-i Sokhta, Sistan," in J.E. van Lohuizen-de Leeuw and J.M.M. Ubaghs, editors, *South Asian Archaeology 1973*, pp. 12–52. Leiden.

Costantini, L. and M. Tosi
  1975 "Methodological proposals for palaeobiological investigations in Iran," in F. Bagherzadeh, editor, *Proceedings of the IIIrd Annual Symposium on Archaeological Research in Iran*. Tehran, Iranian Centre for Archaeological Research. pp. 311–331.

Jux, U. and K.E. Kempf
  in press "Regional geography of Afghan Sistan," in M. Tosi, editor, *Prehistoric Sistan I*. Istituto Italiano per il Medio ed Estremo Oriente (IsMEO) Reports and Memoirs, vol. 18, no. 1. Rome.

Piperno, M. and M. Tosi
  1975   "The graveyard of Shahr-i Sokhta, Iran," *Archaeology,* vol. 26, no. 3, pp. 186-197.

Tosi, M. and M. Piperno
  1973   "Lithic technology behind the ancient lapis lazuli trade," *Expedition,* vol. 16, no. 1, pp. 15-23.

THE CAMEL: ITS DISTRIBUTION AND STATE OF DOMESTICATION IN THE MIDDLE EAST
DURING THE THIRD MILLENNIUM B.C. IN LIGHT OF FINDS FROM SHAHR-I SOKHTA

Bruno Compagnoni, Servizio Geologico d'Italia
Rome
Maurizio Tosi, Istituto Universitario Orientale
Naples

Osteological Finds

Five bones of Camelus sp. were found at Shahr-i Sokhta: the distal portion of a femur, a calcaneum, and three second phalanges (Table 1). Comparison of these specimens with comparable parts from four recent Camelus bactrianus L. and seven recent Camelus dromedarius L. skeletons[1] failed to turn up any characters which could be reliably used to assign the Shahr-i Sokhta bones to one species or the other. A possible exception is the calcaneum as will be detailed below.

The femur (Fig. 1B) comes from provenience LIII.8, a stratum and room that almost certainly dates to structural Phase 6 (ca. 2600-2500 B.C.). The context is a fill deposit in the "House of Foundations." The specimen is the distal portion of a right femur with robust features whose size falls within the range of the reference skeletons (Table 5). All specimens--ancient and modern--are morphologically very similar and no species distinction is possible based on the features of this element.

The calcaneum (Fig. 1A) comes from Room X.8, also dated to Phase 6 and situated in the "House of Foundations." The various dimensions of this piece fall within the range of variation of the comparative specimens independent of species. From a strictly morphological point of view, the shapes of the proximal end (corpus), the distal (articular) end, and the surfaces which articulate with the talus, the lateral malleolus, and the centroquartal all vary widely even within the same species. Our piece is therefore alternatively close to various comparative specimens, with no distinction possible between C. bactrianus and C. dromedarius. Certain metrical characters, however, seem to reoccur more often in one or the other species. Thus, the dimensions of the articular surface with the talus on the substentaculum in the bactrian camel cluster nicely--the depth and width dimensions being almost equal-- with the only exception being specimen B.M. 673a. For the dromedaries, the width of this articular surface tends to be markedly less than its depth with the exception of specimen 52325 (Field Museum) where the opposite is true. Our calcaneum from Shahr-i Sokhta yields depth-width dimensions which are almost identical and thus seems closer to the bactrian condition. A similar situation exists in the relation between the lengths of the distal and proximal ends of the calcaneum.[2] In the bactrian reference specimens as well as in the example from Shahr-i Sokhta, the distal end is less extended than in the dromedary specimens. Exceptions again are B.M. 673a and F.M. 52325. In conclusion, we do not believe that these differences are sufficient to definitely assign the calcaneum, and by association the other finds from Shahr-i Sokhta, to the species C. bactrianus, particularly considering the small number of reference skeletons examined and the existing exceptions for both species. The identification, however, can be considered probable.

The final three camel bones found are all second phalanges: 1) anterior, (probably) left lateral (Fig. 1D); 2) posterior, right medial with a small portion of the distal end missing (Fig. 1C); and 3) posterior, (probably) left medial with the whole distal portion missing. It has been possible to compare these phalanges with those of two bactrian and two dromedary camels from reference collections. The dimensions of the comparative specimens bracket those from Shahr-i Sokhta (Table 5), and furthermore, no constant morphological differences could be discovered to distinguish between C. bactrianus and C. dromedarius or between either and our finds. As far as camels from eastern Iran and Central Asia are concerned, the length dimensions given by Calkin (1970, p. 157) for the phalanges of bactrian camel found in southern Turkmenia are markedly larger than those of our specimen (mm 78 and 79 as opposed to 66.2) and very close to those given by Duerst (1908, pp. 383-384) for the camel phalanges from Anau (74 mm, although not specified whether anterior or posterior).

Both the anterior and the right posterior phalanges come from the fill of the "House of

TABLE 1

Bones of Camelus sp. from Shahr-i Sokhta

Element	no. of specimens rt.	lft.	provenience	period	phase	notes* a	b	c	d
Femur	1		LIII.8	II	6	-	-	-	+
Calcaneum		1	X.8	II	6	-	-	-	-
Phalanx 2 ant.		1	X.3	II-III	5-6	-	-	-	+
Phalanx 2 post.	1		XIII.8	II	6	-	-	-	?
Phalanx 2 post.		1	St.1.9	II	7	-	-	?	-

*notes: a: marks of butchering   +: present
b: marks of intentional working   -: absent
c: marks of teeth   ?: uncertain
d: marks of fire

Foundations" and can be assigned to Phase 6 as a *terminus post quem non*. It is no easy matter to give a specific chronological assignment to the materials coming from this context in spite of the fact that the stratigraphic position can be delimited accurately. At Shahr-i Sokhta, as in most large settlements of the third millennium B.C., the foundations of new constructions were anchored onto the leveled-down walls of the previous construction period. The method used in our case was that of building the foundations of a later building directly onto the floor of an earlier house, practically wedging the new walls between the walls of the earlier rooms. The spaces left after the foundations were built were filled up with inert elastic material. Nothing served this purpose better than debris from one of the heaps of refuse in the vicinity. Thus, between one wall and another we find material that can be dated only to some time earlier than the covering floor. This situation holds for Cuts 3-8 in Rooms X, XII, and LIII which are all part of the Phase 5 foundation filling. On the strength of associated cultural finds, the bones from this fill should come from animals that died before 2500 B.C. The left posterior phalanx from Cut 9 in Street 1 is stratigraphically situated in a layer that should date prior to Cuts 8-10 in the "House of Foundations" and can be assigned to Phase 7 with greater certainty since it is from a microstratified road deposit (Biscione et al. 1974).

Nonosteological Finds

The exceptionally good state of preservation of various forms of organic matter at Shahr-i Sokhta has made it possible to recover further evidence of the camel's presence in protohistoric Sistan. This additional material consists of camel dung and camel hair fibers contained in fragments of cloth.

Dung. The dung find from Shahr-i Sokhta consists of 384.25 grams of dry residue composed of about 600 cylindrical discoids of a dark brown color (Fig. 2B). The discoids vary considerably in weight and size. The smallest ones measure 0.90 cm in diameter and weigh little more than 100 milligrams. The largest ones, often made up of several smaller disks fused together, have a maximum diameter of 2.1 cm and weigh up to 3.0 grams. A number of them appear to have been squashed while still fresh, but the generally excellent state of preservation seems to indicate that they were gathered very soon after excretion.

Fortunately the dung comes from a well-defined stratigraphic context. It was contained in a flat bodied, biconical jar used as a container after having been buried in the floor of Room CLXVI, corresponding to Cut 12. The jar was placed in the northeast corner of the room and only the rim was left sticking out (Fig. 2A). The dry dung occupied about four-fifths of the volume of the jar. Room CLXVI is part of a dwelling situated in the Sector RWD. In spite of the almost complete erosion of its south side, the room can be linked satisfactorily to the surrounding structures [Layer ⑦], all of which are closely associated with Phase 7 (ca. 2700-2600 B.C.). Such a dating overlaps with that estimated for the bones in the East Residential Area.

The identification of the dung was made possible through the kind cooperation of Professor F. Baschieri, Deputy Director of the Rome Zoo. The Shahr-i Sokhta discoids were compared, in March

FIGURE 1. *Bones of Camelus sp. from Shahr-i Sokhta (scale in centimeters is approximate).*
  A. *left calcaneum [provenience X.8]*
    1. *medial view*
    2. *dorsal view*
    3. *lateral view*
  B. *right femur [LIII.8], cranial view*
  C. *posterior phalanx 2 [XIF.8.XIII] dorsal view*
  D. *anterior phalanx 2 [X.3] dorsal view*

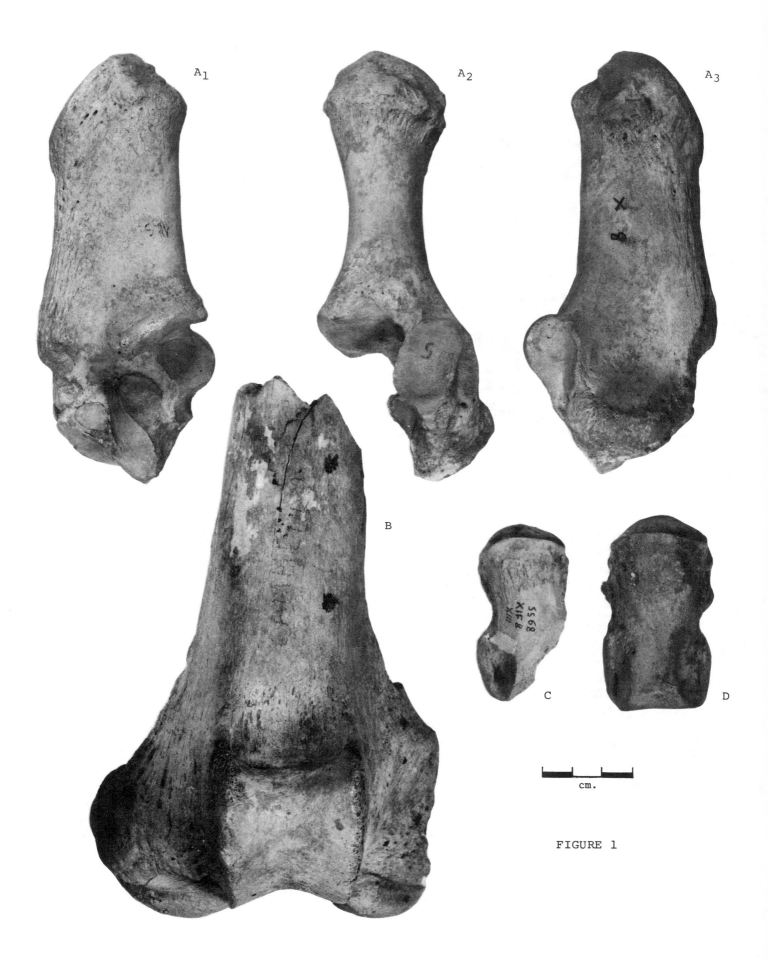

FIGURE 1

FIGURE 2A

FIGURE 2C

FIGURE 2D

FIGURE 2B

FIGURE 3

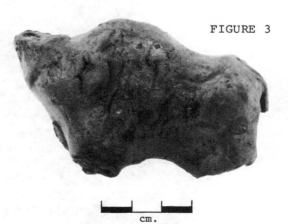

FIGURE 2. *Camel dung from Shahr-i Sokhta and comparative material.*
  A. *Shahr-i Sokhta excavations showing the position of Room CLXVI with buried jar close to the eastern wall. The room is part of a dwelling complex dated to Phase 7.*
  B. *Dry camel dung from inside buried jar of Room CLXVI*
  C. *Dried camel dung pellets from Shahr-i Sokhta*
  D. *Camel dung from the Rome Zoo collected in May 1972*

FIGURE 3. *Clay animal figurine from Shahr-i Sokhta Room LXIX.10, associated with destruction level at the end of Phase 7 [inventory number 4501].*

1972, with material gathered in the cages of C. dromedarius and C. bactrianus. Our samples were found to be identical with those from a young (10 month) female C. dromedarius fed on dry fodder (Fig. 2C-D). The identification, however, is valid at the genus level only because the sole differences noted in the dung of the two species were due to age and feed. The Shahr-i Sokhta find is thus the dung of a young camel fed on dry grass. Evidently the dung had been gathered over a period of a few days.

The finding of dung at Shahr-i Sokhta allows us to establish two points: 1) in Sistan, the camel was being used as early as about 2700 B.C. for more than just meat and hides, thus implying a more complete utilization of the animal; 2) young animals were evidently being kept in the vicinity of the community with the result that their dung could be gathered soon after excretion.

Hair. Camel hair was used to spin thread at Shahr-i Sokhta. Evidence for this practice comes from the microscopic analysis of a few fragments of material found under the hardened salt surface which sealed an ancient storeroom. In the two cases where the camel fibers have been discovered, the hair was found to have been woven together with a much larger quantity of fiber from goat or sheep. Microscopic examination revealed necrotization of the fiber teguments which made it difficult to distinguish between goat and sheep hair. The fibers of camel hair, however, are easier to pick out because of their greater size (6 times larger on the average), the compactness of their outer tegument, and the existence of a supply canal that is four-fifths of the total diameter (Fig. 4A). According to the literature at our disposal, it does not appear possible to distinguish between the various Camelus species on the basis of a superficial examination of hair fibers.

The first of the finds analyzed (inventory number 4234) is a fragment of a strip of yellowed fibers 10 cm long and 1.4 cm wide. The weave consists of seven bundles of warp stretched by the weft according to the combination 3-1-3 (Fig. 4B). The strip was found in Room CCXIIIa ② which is the under-stairs storeroom on the east side of the "House of the Pit." This room was filled at the end of Phase 6 with a load of refuse consisting mainly of organic matter.

The other piece with camel fibers is one of the fragments of fabric found in RYL 2. These fragments all consist of smooth materials and are of a simple weave which is rather loose because, perhaps, of the thickness of the threads and the instability of the spun fibers. The context from which these fragments come is chronologically linked to the Phase 5 structures which represent the upper layer in Sector RYL. These pieces are in no way different from the other cloth fragments of Period II found in other residential areas and in a number of graves. They all are of a type having a smooth warp and a simple weft pattern. The yarn used varies considerably in thickness and there are thus a few pieces which are particularly soft and light. The only fibers found so far are animal in origin.

Figural Representations. Having established the presence of the camel in the faunal complex of Shahr-i Sokhta and suggested the presence of a complex pattern of economic utilization, it is logical to suppose that the animal was also occasionally depicted on pottery, as figurines, or by other means.

The representation of animals is frequent in the material culture of prehistoric Iran, both on pottery and as coroplastic art. The gradual geometrization of painted decoration starting after 3500 B.C. and its subsequent abandonment in the second half of the third millennium throughout most of the plateau and basins of Iran means that only in the coroplastic art is it possible to find representations of camel in the period of interest to us. Unbaked clay and terracotta figurines are common objects at Shahr-i Sokhta during Periods I-III. The care taken in the modeling varies widely, especially in the figurines made of unbaked clay. Many of these were later deformed by pressure and erosion and nearly all are mutilated. In spite of this poor state of preservation, it has been possible to compile Table 2 from about 1200 figurines found in layers of Phases 8-5. Comparison of the percentages of the representations with those of the osteological finds reveals no correlation. The use of a given genus in the city's economy does not seem to have been the principle on which the modeling of the figurines was based, except, perhaps, in the case of Bos where a 21.50% presence among the osteological material corresponds to the highest incidence among the representations.

Figurines of the camel are still absent from the Shahr-i Sokhta assemblage. Inventory number 4501 from provenience LXIX.10 has given rise to heated discussion among the members of our mission but we have been compelled to classify the find as unidentifiable. It is a clay figurine 5.05 cm long and 4.2 cm high with the head and legs missing. Features which suggest that it might represent a camel are the shape and position of the large hump, the upward rising nature of the cylindrical neck, and the short tail. Evidence to the contrary are the treatment given to the hindquarters, the existence of a "bony" apex corresponding to the position of the pelvis, and, in general, the abundance and variety of zebu figurines to which no. 4501 is comparable. A photograph is included here as critical documentation (Fig. 3).

General Considerations

Having reviewed the evidence for the camel at Shahr-i Sokhta, the question that is raised now is what contribution do these new finds make toward clarifying the problems of the domestication of the camel and the distribution of the two species in the area of the Middle East?

A vast literature exists on both subjects, begun, practically speaking, when J.U. Duerst wrote his essay on the animal remains from Anau (1908). There, only vertebrae and phalanges were found and their assignment to the species C. bactrianus was made primarily on the basis of the geographic position of the site. Zeuner (1963, p. 359) subscribed to Duerst's identification because of the large size of the phalanges. Doubtless under the influence of the methodological approach laid down by R. Pumpelly and E. Huntington for all the research carried out by the American expedition to Turkestan, Duerst saw the presence of the camel in a chalcolithic settlement of the southern steppes as confirmation of the close relationship between the domestication of the horse and camel and the process of desertization of the southern Palaearctic strip. Indeed, the bactrian camel was certainly closely associated with the open land and severe continental climate that characterized the Eurasian steppes during the postglacial age. We find the camel even in the Ukraine associated with wild horse and saiga antelope in the Tripolje settlements.

It is not worthwhile to repeat here the intricacies of the discussion in progress on the distribution of the bactrian camel on the Iranian plateau since the various arguments have been recently summed up by Bulliet (1975). The main points of the problem revolve around a basic discrepancy: the bactrian camel does not exist today west of the Amu dar'ja, while all the prehistoric osteological and representational finds

FIGURE 4. *Textile from Shahr-i Sokhta.*
    A. *Magnified view of wool fibers from textile [inventory number 4234] dated to Phase 6. The thick camel hair is easily distinguishable from those of sheep/goat. (Magnification ca. 18.5 times.)*
    B. *Wool fragment from Shahr-i Sokhta Room CCXIII.a [inventory number 4234]*

FIGURE 5. *Limestone orthostats from Umm an-Nar (Abu Dhabi, United Arab Emirates) from the ring-wall of Grave III [by courtesy of Karen Frifelt].*

TABLE 2

Zoomorphic figurines from Shahr-i Sokhta
[Phases 8-5: ca. 2900-2500 B.C.]

Genera/species	unbaked clay specimens	terracotta specimens	totals	% of identified specimens	% of total specimens
Canis sp.	57	3	60	5.32	4.66
Felis sp.	5	3	8	0.71	0.62
Equus sp.	16	0	16	1.42	1.24
Sus scrofa L.	7	47	54	4.79	4.19
Gazella sp.	2	1	3	0.27	0.23
Bos indicus L.	661	69	730	64.72	56.68
Bos taurus L.	16	0	16	1.42	1.24
Bos sp.	161	31	192	17.02	14.91
total *Bos* sp.	838	100	938	83.16	72.83
Capra sp.	18	1	19	1.68	1.48
Ovis sp.	21	1	22	1.95	1.71
AVES	5	3	8	0.71	0.62
unidentified	76	84	160	--	12.42
TOTALS	1045	243	1288	100.01	100.00

[Information kindly provided by Dr. Alfredo Coppa, Istituto di Antropologia, Rome]

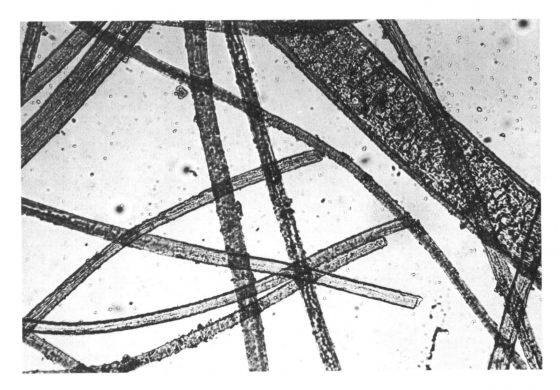

FIGURE 4A

FIGURE 5

FIGURE 4B

in Iran point to its presence as far west as Kashan and as far south as the Makran (Fig. 6). Table 3 summarizes the evidence for the period corresponding to the expansion of the East Iranian proto-urban cultures, tentatively identifying the Sistan finds as C. bactrianus. In all there are only a small number of finds--barely twenty-two bones and less than ten representations, some of which are doubtful. Even such a sporadic distribution, however, can provide us with indications of considerable interest. First and foremost, the finds from Shahr-i Sokhta and those from the Soviet excavations in southern Turkmenia were not available to Zeuner at the time he wrote his monumental A History of Domesticated Animals (1963) and thus represent a completely new contribution to the discussion as crystallized by this work.

In addition, the occurrences of C. bactrianus in eastern Iran suggest two new considerations. Reading Table 3 in a horizontal direction it is easy to see what the geographic area of distribution of the finds is. The sites of Turkmenia, the plain of Gorgan, the oasis of Kashan, the Bampur valley and Sistan form a circle around the "region of the East Iranian basins"--the Khavir/Lut--a large residue of the tertiary Tethys lying in geological and environmental contrast to the Iranian and Baluchi plateaus running along it. The territory is quite desertic in the central part, but gradually becomes richer in vegetation and animal species as one approaches the piedmont borders. The faunal association in this environment is characterized by running gregarious mammals such as the goitered gazelle (Gazella subgutturosa) and the onager (Equus hemionus), both of which are preyed upon by typical running carnivores such as the leopard (Felis pardus L.) and the cheetah (Acinonyx jubatus Schreber) (Misonne 1959; Lay 1967; Firouz 1974). C. bactrianus is certainly a species suited to this sort of environment with its dry continental climate, its altitude of 500-1000 meters, its large spaces of steppe, and its bush-grass vegetation with a true pabular covering between February and May. The geographic distribution of the archaeological finds thus suggests that the area of distribution of wild C. bactrianus should be extended to the south of the 45th parallel of latitude which roughly marks the southernmost boundary of the cold Eurasian steppes, thereby including the entire Turanic steppes and the East Iranian basin region (Fig. 6). To the south, the zone would be delimited by the hydrographic complex of the Halil Rud, Jaz-i Murian, and Bampur/Damin drainages, an area which contained substantial numbers of settlements during the third millennium B.C. (Tosi 1974a). It is from Tomb E at Khurāb that there comes an axe head with a representation of a camel which, on the basis of technical examination, is interpreted to be C. bactrianus (Lamberg-Karlovsky 1969, with further considerations in During Caspers 1971). Lying between Khurāb and southern Turkmenia there is a natural bridge in the form of the intramontane plateau of the Kuh-i Birjand, running along the 60th meridian of longitude. This is true bushy steppe region with thickly growing halophyte species and is the present-day habitat of Gazella subgutturosa. Small herds of C. bactrianus could easily have moved over this

TABLE 3

Evidence for the camel in the third millennium in eastern Iran and southern Turkmenia

Period	Southern Turkmenia[1]	Gorgan Plain[2]	Oasis of Kashan[3]	Sistan: Shahr-i Sokhta	Makran
3500-3000	Anau II: b		Siyalk III_4: r		
3000-2500	Šor depe: b Čong depe: b Hapuz depe: b			II: b,h,d	
2500-2000	Ulug depe: r Altyn depe: b,r Namazga depe: b,r	Shah tepe: b		II-III: b,h	Khurāb E: r

b = bones   r = representation   h = hair   d = dung

Notes: 1. Masson and Sarianidi 1972, p. 109
2. Amschler 1940
3. Ghirshman 1938, vol. I, pl. LXXIX, A2

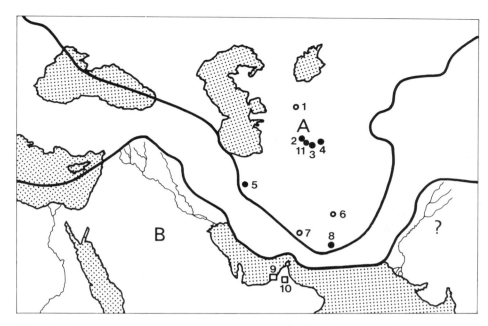

FIGURE 6. *Map of the reconstructed areas of distribution of Camelus bactrianus (A) and Camelus dromedarius (B) compiled from archaeological data.*

1. Tuiek-kičidžik (information kindly supplied by A.V. Vinogradov)
2. Anau
3. Altyn depe
4. Čong depe
5. Siyalk
6. Shahr-i Sokhta
7. Tepe Yahya
8. Khurāb
9. Umm an-Nar
10. Hili
11. Šor depe

● Camelus bactrianus
○ Camelus cf. bactrianus
□ Camelus dromedarius

diaphragm spending the winter months in the playas, thus being sporadically within range of the Shahr-i Sokhta hunters.

The southern border of the area thus delimited coincides with the southwest boundary of the Eurasian plate and the northeast boundary of the African plate. This barrier, represented by the Zagros Mountains, rose gradually in the course of the Tertiary era and may have created conditions suitable for the separation of the two camelid species with respect to their assumed Pliocene forebear, Camelus sivalensis Falconer (Zeuner 1963, p. 340).

The alluvial plain of Sind, however, may have been a territorial appendix of the area of distribution of C. dromedarius. B. Prashad (1936, pp. 58-59) has recognized the presence of the dromedary among the osteological finds from Harappa, and Sewell and Guha (1931, p. 660) have done the same for Mohenjo Daro. There has been much discussion as to whether these presumably third millennium materials were, in fact, intrusive into their respective deposits with the proposition seemingly supported by the absence of the camel in the rich zoomorphic repertoire of proto-Indian glyptics (Rizzi 1970; Conrad 1966, p. 61). The subtropical desert environment of Sind, however, seems to be quite suitable for C. dromedarius, and proto-Indian glyptics have been found not to have represented numerous other genera that were present in the osteological material from Harappa and Mohenjo Daro including Ovis, Equus, and Sus. In addition, recent excavations by the French Archaeological Mission at Pirak on the Kacchi Plain (near the Bolan Pass) have brought to light a large number of small terracotta figurines representing not dromedary but bactrian camels with clearly emphasized humps (Jarrige and Enault 1976, p. 44, pl. XIII, 7; XX, 3). The date of Pirak is between 1800 and 900 B.C. It is possible, therefore, that the identification of C. dromedarius by Prashad and others might be open to question. In any event, the areas of distribution of the bactrian camel and of the dromedary were limited by local phytogeographic and climatic conditions rather than by strictly latitudinal considerations. It is logical, therefore, that there should be an overlapping strip lying roughly between 27° and 38° north latitude, in which, owing to the action of man, the dromedary has become the predominant camel.

In conclusion, in the third millennium all archaeological evidence points to the fact that the dromedary occupied an area lying between Oman (Fig. 5) and Sind to the east (Thorvildsen 1964, Figs. 7-8; Tosi 1974b, pp. 162-163 and Fig. 12) and North Africa and Palestine to the west (Free 1944; Mikesell 1955, pp. 236-238; Zeuner 1963, pp. 349-352; Epstein 1971, pp. 580-584; Ripinsky 1975 and n.d.). In Mesopotamia, as is to be expected on the basis of existing environmental conditions, all of the rather rare representations known to be of third millennium date are of dromedaries.

Let us now consider the second new point suggested by the East Iranian finds. Reading Table 3 in a vertical (chronological) fashion, it can be seen at once that the most complete sequence of finds over the third millennium occurs in southern Turkmenia. As was stressed by Calkin (1970, p. 156), the camel appears at a relatively late stage in the southern Turkmenia sequence. No bones are found in the well-explored Neolithic facies (5500-4500 B.C.), characterized by the settlement at Džejtun where hunting activity accounted for 65 percent of the faunal remains. Camel appears in the fourth millennium (Čong depe) and establishes itself during the third millennium simultaneously with a sharp increase in the human population. During this later period, hunting was on the wane and the remains of wild animals are restricted to outlying settlements like Šor-depe (58.2 percent) and to the poorer quarters of the towns (Ermolova 1968, p. 49; 1970, p. 207). According to Calkin, the distribution data would thus indicate that the increasing presence of the camel in the faunal record is an index of its cohabitation with man. Chronologically speaking, the southern Turkmenian evidence thus supports the data from Sistan in the sense that both may be interpreted as evidence of an ongoing process of domestication.

The low percentage of camel bones in a faunal assemblage can hardly be used to argue against an hypothesis of ongoing domestication. The studies made by N.M. Ermolova on the fauna from Šehr-Islam (9th-14th century A.D.) in the greater Merv area which lies in the center of a vast Central Asian caravan trade network have shown that camel bones account for 0.4 percent of the total (Ermolova 1970, p. 215). Because the primary function of the camel is an extra-urban one often connected with endemic nomadism, it is not surprising that the animal is always poorly represented in urban settlements.

In the case of the camel, every inquiry into the process of its domestication comes up against a truly insurmountable barrier--our ignorance of the behavior, habitat, and osteometry of the wild species. The studies carried out by Ivor Montagu seem to demonstrate that the Khautagai (C. bactrianus L.) found in the center-south regions of Outer Mongolia (Nomin Gobi) are indeed wild (Bannikov 1958, pp. 156-159; Montagu 1965; Namnandorj 1970, p. 15). Zeuner mentions studies in progress and a film made by the Mongols Jidig and Damdin which was shown in London in 1956. Unfortunately, no more detailed studies are known and the habitat and behavior of C. bactrianus still remains to be reconstructed deductively. The animal probably lived in small herds composed mainly of females and young animals led by an adult male. The present-day habits of the animals suggest that it must have been a fairly easy matter to corral the small herd once the adult males had been eliminated. Young animals and females could easily have been kept near a settlement and their hides, meat, milk, dung, urine, and hair used without thus bringing about any physiological transformation of the animal or any socioeconomic change in attendant human activities.

The particular connotation of the term "domestication" must vary species by species to the extent that it implies a form of socio-behavioral contract between man and individual members of a gregarious species. Domestication can be said to have taken place only when the process of animal keeping has, to some extent, resulted in changing morphological and behavioral characteristics of the animals and also in altering the socioeconomic structure of the human community. Selection of particular characteristics will be determined by the nature of the interactions between species. Once domestication is conceived of as a selective diachronic process, the domestication of the camel can be said to have <u>begun</u> as soon as man was in a position to exploit continuously the physiological processes of the species, but to have been <u>achieved</u> only when the

TABLE 4: Evolution of Bedouin society, after Dostal 1958

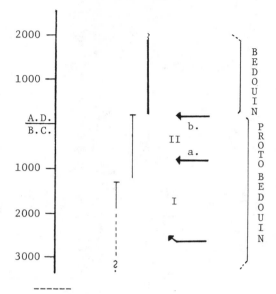

I. Contacts with urban civilizations.
IIa. First influences of horse riding techniques.
IIb. Introduction of the bow saddle for camels; development of war techniques with the dromedary.

animal came to be used as a beast of burden and for riding, thus opening up new territories for human occupation in the subtropical desertic strip and forcing the human groups who exploited the animal to become specialized in the socioeconomic dimension of their culture. Such is the logic underlying the development of Bedouin society which Walter Dostal deduced from the few elements available as early as 1958 (p. 13). In his brief survey, which is basically a study of riding techniques, he outlined the various phases in the evolution of Bedouin society, seeing its development as a function of the development of riding animals and of the control of the dromedary in Arabia and North Africa (Table 4).

The concept of a "proto-Bedouin stage" introduced by Dostal seems to be in perfect accord with the archaeological evidence from protohistoric eastern Iran even though, there, the species in question seems to have been C. bactrianus. The term Bedouin is applied by Dostal to members of nomadic tribes of camel raisers whose animals were used as mounts in war and peace and who possessed large breeding herds. In their extremely high degree of socioeconomic specialization, they reveal that total accord with the gregarious species and its environment that we have termed "domestication."

The discoveries of the Danish Archaeological Mission on Umm an-Nar off the coast of Abu Dhabi, where representations in relief on tombstones and osteological finds of camel are dated as early as 2500 B.C., have revealed a convergence between Iran and Oman, in the more highly evolved farming communities, in the domestication process of C. bactrianus and C. dromedarius. For camel as a beast of burden, the earliest documentation we have dates to the beginning

TABLE 5

Measurements of camel bones from Shahr-i Sokhta and comparative collections
(in millimeters)

Femur	a	f	g	i	j	n	o	p
1. max. distal width	121.0	103.0	125.0	105.0	107.5	122.0	135.0	128.0
2. max. A-P distal diameter	115.0	105.0	123.0	114.0	108.0	130.0	133.0	--
3. length of trochlea	75.0	66.0	76.5	73.5	69.0	83.0	84.0	--
4. min. width of trochlea	37.3	33.5	40.0	32.0	35.0	37.5	39.0	35.0

Calcaneum	b	f	g	h	i	k	l	m	n	o	q
1. max. length	150.0	134.0	148.0	150.0	137.0	130.0	147.0	123.0	155.0	165.0	146.5
2. min. width of corpus	23.1	18.2	29.5	24.2	22.1	20.8	22.0	20.0	26.4	23.2	22.3
3. max. dorsal-plantar depth	66.6	55.0	70.0	76.5	60.0	61.2	66.0	53.2	71.0	78.0	67.3
4. length of proximal end	97.0	84.0	98.0	93.5	81.0	79.5	92.2	78.6	97.2	109.0	94.5
5. length of distal end	59.0	54.0	55.0	58.0	61.5	55.0	61.0	51.2	63.3	62.4	54.5
6. index: 5 x 100/4	60.8	64.3	56.1	62.0	75.9	69.2	66.2	65.1	65.1	57.2	57.5
7. depth of articular surface with talus on sustentaculum	30.5	27.5	36.3	37.0	36.0	34.0	37.0	32.0	38.5	33.5	33.3
8. width of articular surface with talus on sustentaculum	30.0	27.0	31.3	34.0	29.5	28.0	32.0	27.5	34.0	37.0	
9. index: 8 x 100/7	98.4	98.2	86.2	91.9	81.9	82.4	86.5	85.9	88.3	110.4	

	Phalanx 2, anterior					Phalanx 2, posterior					
	c	f	h	i	o	d	e	f	h	i	o
1. max. length	66.2	60.3	69.0	64.0	76.0	(60.5)	--	54.8	64.5	57.2	69.8
2. min. width	29.0	29.6	26.2	26.7	30.6	(24.0)	23.5	25.2	25.0	26.2	27.3
3. max. proximal width	34.6	33.0	37.0	33.8	41.5	30.1	29.8	29.5	34.0	29.8	38.2
4. max. proximal depth	27.6	25.0	28.7	26.2	33.0	24.2	--	22.0	25.5	23.2	30.0
5. max. distal width	35,5	35.0	36.3	34.0	43.0						
6. max. distal depth	18.2	17.3	19.5	17.1	23.0						

a = femur from Shahr-i Sokhta LIII.8
b = calcaneum from Shahr-i Sokhta X.8
c = phalanx 2, anterior from Shahr-i Sokhta XIF.3.X
d = phalanx 2, posterior from Shahr-i Sokhta XIF.8.XII
e = phalanx 2, posterior from Shahr-i Sokhta St.1.9
f = *Camelus bactrianus*, Field Museum 18847
g = *Camelus bactrianus*, British Museum 673.a
h = *Camelus bactrianus*, British Museum 1947.10.21.5
i = *Camelus dromedarius*, Museo Civico di Zoologia di Roma 21395-5/6
j = *Camelus dromedarius*, Museum of Comparative Zoology, Harvard 51320
k = *Camelus dromedarius*, Museum of Comparative Zoology, Harvard 51353
l = *Camelus dromedarius*, Museum of Comparative Zoology, Harvard 51352
m = *Camelus dromedarius*, Museum of Comparative Zoology, Harvard 16891
n = *Camelus dromedarius*, Field Museum 8411
o = *Camelus dromedarius*, Field Museum 52325
p = *Camelus bactrianus domesticus*, femur, measurements deduced from illustration 1a in plate 28 of Gromova 1950.
q = *Camelus bactrianus ferus*, calcaneum, measurements deduced from illustration 46 of Gromova 1960

of the second millennium B.C. with the representation of a camel on a plaque from Tell Asmar in Mesopotamia (Epstein 1971, p. 567) and the figurines of C. bactrianus from Altyn depe and Ulug depe in South Turkmenia (Calkin 1970, Fig. 4).

The domestication of the camel was thus a slow process of mutual assimilation that reached the height of its specialization only on the Eurasian steppe and in the subtropical deserts under conditions not unlike those of the natural habitats of each species in the wild state. Because of the sporadic presence of small wild herds, the camel was perhaps always a rare animal until it was possible to increase its numbers by means of gradual symbiosis with the human community. We believe that it is this early stage that we have documented at Shahr-i Sokhta during the first half of the third millennium B.C.

Notes

1. For Camelus bactrianus: Field Museum of Natural History no. 18847 and 18848 (juvenile); British Museum (Natural History) no. 1947.10.21.5 and 673.a; for Camelus dromedarius: Museo Civico di Zoologia di Roma no. 21395-5/6 (female); Museum of Comparative Zoology, Harvard Univ. no. 16891, 51320, 51352, 51353; Field Museum of Natural History no. 8411 and 52325 (castrate).

2. The distal end has been measured between the coracoid process and the beak; the proximal end has been measured between the tuberosity and the beak.

References

Amschler, J.W.
1940   "Tierreste der Ausgrabungen von dem 'Grossen Königshügel' Shah Tepé, in Nord-Iran," in *Reports from the Scientific Expedition to the Northwestern Provinces of China under the Leadership of Dr. Sven Hedin - The Sino-Swedish Expedition.* Publication 9, VII. Archaeology 4, pp. 35-129. Stockholm.

Bannikov, A.G.
1958   "Distribution géografique et biologique du cheval sauvage et du chameau de Mongolie *(Equus przewalskii et Camelus bactrianus),*" *Mammalia,* vol. 22, no. 1, p. 152-160.

Biscione, R., G.M. Bulgarelli, L. Costantini, M. Piperno, M. Tosi
1974   "Archaeological discoveries and methodological problems in the excavations of Shahr-i Sokhta, Sistan," in J.E. van Lohuizen-de Leeuw and J.M.M. Ubaghs, editors, *South Asian Archaeology 1973,* pp. 12-52. Leiden.

Bulliet, R.W.
1975   *The Camel and the Wheel.* Cambridge, Mass.

Calkin, V.I.
1970   "Drevnejšie domašnie životnye Srednej Azii, Soobščenie I," *Bjulleten' Moskovskogo Obščestva Ispytatelej Prirody - Otdel Biologičeskij,* vol. 75, no. 1, pp. 145-159.

Conrad, R.
1966   "Die Haustiere in den frühen Kulturen Indiens." Dissertation. München.

Dostal, W.
1958   "Zur Frage der Entwicklung des Beduinentums," *Archiv für Völkerkunde,* vol. 13, pp. 1-14.

Duerst, J.U.
1908   "Animal remains from the excavation at Anau and the horse of Anau in its relation to the history and to the races of domesticated horses," in R. Pumpelly, editor, *Explorations in Turkestan, Expedition of 1904,* pp. 341-442. Washington.

During Caspers, E.C.L.
1971   "La Hachette trouée de la Sépolture E de Khurāb, dans le Balouchistan persan: Examen retrospectif," *Iranica Antiqua,* vol. 9, pp. 60-64.

Epstein, H.
1971   *The Origin of Domestic Animals of Africa,* vol. 2, pp. 545-584. New York.

Ermolova, N.M.
1968   "Kostnye ostatki mlekopitajuščih iz poselenij epohi eneolita i bronzy Južnogo Turkmenistana," *Karakumskie Drevnosti,* vol. 1, pp. 48-53.
1970   "Novye materialy po izučeniju ostatkov mlekopitajuščih iz drevnih poselenij Turkmenii," *Karakumskie Drevnosti,* vol. 3, pp. 205-232.

Firouz, E.
1974   *Environment Iran.* Tehran.

Free, J.P.
1944   "Abraham's Camel," *Journal of Near Eastern Studies,* vol. 3, pp. 187-197.

Ghirshman, R.
1938   *Fouilles de Sialk, près de Kashan,* Paris.

Gromova, V.I.
1950   *Opretelitel' mlekopitajuščih SSSR po kostjam skeleta,* vol. 1. Trudy Komissii po izučeniju cetvierticnogo perioda, vol. 9. Moskva-Leningrad.
1960   *Opretelitel' mlekopitajuščih SSSR po kostjam skeleta,* vol. 2. Trudy Komissii po izučeniju cetvierticnogo perioda, vol. 16. Moskva-Leningrad.

Jarrige, J.-F. and J.-F. Enault
1976   "Fouilles de Pirak - Baluchistan," *Ars Asiatiques,* vol. 32, pp. 29-48.

Lamberg-Karlovsky, C.C.
1969   "Further notes on the shaft-hole pick-axe from Khurāb, Makran," with an appendix by H.N. Lechtman, *Iran,* vol. 7, pp. 163-168.

Lay, D.M.
1967   *A Study of the Mammals of Iran, resulting from the Street Expedition of 1962-63.* Fieldiana: Zoology, 54, Chicago.

Masson, V.M. and V.I. Sarianidi
1972   *Central Asia: Turkmenia before the Achaemenids.* London.

Mikesell, M.W.
1955   "Notes on the dispersal of the dromedary," *Southwestern Journal of Anthropology,* vol. 9, pp. 231-245.

Misonne, X.
1959 *Analyse zoogéographique des mammifères de l'Iran*. Mémoires de l'Institut Royal des Sciences Naturelles de Belgique, deuxième série, fasc. 59. Bruxelles.

Montagu, I.
1965 "Communication on the current survival in Mongolia of the wild horse *(Equus przewalski)*, wild camel *(Camelus bactrianus ferus)*, and wild ass *(Equus hemionus)*," *Proceedings of the Zoological Society*, vol. 144, pp. 425-428.

Namnandorj, O.
1970 *Conservation and Wild Life in Mongolia*. Miami, Field Research Projects.

Prashad, B.
1936 *Animal Remains from Harappa*. Memoirs of the Archaeological Survey of India, vol. 51. Delhi.

Ripinsky, M.M.
1975 "The camel in ancient Arabia," *Antiquity*, vol. 69, pp. 295-298.
n.d. The occurrence of the camel in the ancient Egypt and the Sahara, mimeographed.

Rizzi, R.
1970 "Tipologia e stile dei sigilli della Valle dell'Indo." Tesi di Laurea per il Conseguimento del titolo Dottore in Lettere. Rome.

Sewell, R.B.S. and B.S. Guha
1931 "Zoological remains," in J. Marshall, *Mohenjo Daro and the Indus Civilization*, vol. 2, pp. 649-673. London.

Thorvildsen, K.
1964 "Gravrøser på Umm an-Nar," *KUML*, pp. 191-219.

Tosi, M.
1974a "Bampur: a problem of isolation," *East and West*, vol. 24, pp. 29-50.
1974b "Some data for the study of prehistoric cultural areas on the Persian Gulf," *Proceedings of the Arabian Seminar*, vol. 4, pp. 145-171.

Zeuner, F.E.
1963 *A History of Domesticated Animals*. London.

THE BONE REMAINS OF EQUUS HEMIONUS FROM SHAHR-I SOKHTA

Bruno Compagnoni, Servizio Geologico d'Italia
Rome

Equus (Asinus) hemionus Pallas, 1775

Most students of equid taxonomy consider asiatic wild half-asses a species of the genus Equus while they differ whether to assign them to a distinct subgenus Hemionus Stehlin e Graziosi 1935 or to include them in the subgenus Asinus Gray 1824. In a recent review of these questions, Groves and Mazák (1967) have presented evidence for the affinities between asiatic and african wild asses, with the result that there is a return to the practice of including them either in the genus Asinus (Groves and Mazák 1967) or in the subgenus Asinus (Groves 1974; Turnbull and Reed 1974). In a similar manner, the number of species into which asiatic half-asses should be divided is a subject of controversy. Some consider them a single species, Equus hemionus Pallas 1775 (e.g., Zeuner 1963; Grzimek 1973; Turnbull and Reed 1974), others as two species (Groves and Mazák 1967; Groves 1974), and others as four species (Bourdelle 1955; Azzaroli 1966), each with numerous subspecies. Without entering into complicated taxonomic questions which are difficult to resolve at present, I think it preferable here to refer to the asiatic wild half-ass as a single species, Equus hemionus Pallas 1775, with a rather large range of variation.

Twenty-one cranial and post-cranial specimens of Equus have been recovered from third millennium levels at Shahr-i Sokhta (Table 1), some bearing traces of burning but none with signs of butchering. Comparison of these remains with numerous modern specimens of Equus hemionus, subspecies: hemionus, kulan, khur, onager, kiang, and hemippus (Table 2), has permitted identification of the Shahr-i Sokhta material as Equus hemionus and possibly as the subspecies E. h. onager Boddaert 1875 (following Misonne [1959] who includes central-eastern Iran in the range of this subspecies).

The Remains from Shahr-i Sokhta

As far as the chronological distribution of the Shahr-i Sokhta equid material is concerned, Table 1 clearly shows that these remains are predominant in the layers of Period I, even though four of them are teeth from the same provenience (CX. 24) and probably from the same animal. Provenience XX.23-24 is a large pile of animal bones from the filling under the earliest floor of the "House of the Stairs." Proveniences CI.22 and CIV.29 are distinct and well-defined in their sequential order. Although the specimens from Period I (ca. 3200-2700 B.C.) do cluster in certain areas, Equus hemionus is quite obviously most common during this period, especially given the fact that the area excavated is only one-twentieth of that for Period II.

Occipital (Fig. 2G). This specimen comes from a juvenile individual (the suture between the exoccipitals corresponding to the nuchal tuberosity is not completely knitted) and lacks the central part of the external occipital protuberance and the extremities of the jugal apophyses.

Except for its greater height, the considerable morphological similarity between the Shahr-i Sokhta occipital and the E. h. kiang specimens B.M. 1879.11.21.194, B.M. 1851.7.16.4, the E. h. hemionus specimen B.M. 12.5.7.1 --the only comparative material with an unworn M^2 and unerupted M^3, with small lateral extensions of the nuchal crests (excepting B.M. 1851.7.16.4), and with large sized nuchal tuberosities and consequent thickening of the upper edge of the foramen magnum--seems to indicate that the Shahr-i Sokhta specimen must have belonged to an individual aged about two and one-half years. The adult half-ass specimens have highly expanded nuchal crests and nuchal tuberosities that, while fairly high, are always thinner than in the Shahr-i Sokhta piece. Our occipital is fairly high (inion-basion distance = ca. 98 mm) and is exceeded only by a very old kiang specimen (B.M. 1851.7.16.5, aged 22 years, with an inion-basion distance of 99 mm). The Shahr-i Sokhta specimen is also relatively narrow with the index (= minimum width x 100/inion-basion distance) of 54.5 greater only than that of E. h. onager (B.M. 1966.3.14.1) and of a few of the E. h. hemionus skulls, while the specimens of E. h. kiang, E. h. khur, and E. h. hemippus are all relatively wider, the first two in an absolute sense as well (Table 3).

In the graph, Figure 1, the inion-basion distances are plotted on the abscissa and the smallest width on the ordinate both for the Shahr-i Sokhta occipital and for the comparative material.

106 Approaches to Faunal Analysis

TABLE 1

Bone remains of Equus hemionus from Shahr-i Sokhta

(2)	Element:	no. of specimens rt.	lft.	Provenience	Period	Phase	Notes[1] a	b	c	d
	teeth: I_1	1		CX.24	I	9	-	-	-	?
	I_2	1		CX.24	I	9	-	-	-	?
	I_3	1		CX.24	I	9	-	-	-	?
	upper ?PM		1	XX.27	I	10	-	-	-	+
	$?M^2$		1	RDC.2	IV	0	-	-	-	+
	M_3	1		CX.24	I	9	-	-	-	?
aa.	occipital		1	XX.21	I	9	-	-	-	?
bb.	squamous temporal		1	CIV.29	I	9-10	-	-	-	-
cc.	scapula		1	s.c.(3)	-	-	-	-	-	+
hh.	metacarpal		1	X.8	II	6	-	-	-	?
jj.	phalanx 1 anterior	1		XX.22	I	9	-	-	?	+
dd.	pelvis	1		XX.23-24	I	9	-	-	-	?
	pelvis		1	s.c.	-	-	-	?	?	?
ee.	femur	1		s.c.	-	-	-	-	-	?
ff.	patella		1	s.c.	-	-	-	-	-	?
gg.	calcaneum	1		s.c.	-	-	-	-	?	+
ii.	metatarsal (juv.)		1	XX.26	I	9	-	-	-	?
kk.	phalanx 1 posterior		1	XII.4	II-III	5	-	-	-	+
ll.+ mm.	phalanx 3 posterior	2		s.c.	-	-	-	-	-	?
nn.	distal sesamoid	1		CI.22	I	9	-	-	-	?

1) Notes: a = marks of butchering
 b = marks of intentional working
 c = marks of gnawing
 d = marks of burning
 + = present
 - = absent
 ? = uncertain
2) identification letters used in Tables 3 and 4.
3) s.c. = undefined provenience

In spite of the small numbers of E. h. onager and E. h. khur specimens available, the following statements can be made: 1) the Shahr-i Sokhta specimen falls outside of the range of the other forms but lies nearest to those of E. h. hemionus and E. h. kiang; 2) the ranges of two last mentioned subspecies overlap; and, above all, 3) while the indices calculated for E. h. onager, E. h. khur, and E. h. kiang show a tendency to decrease with increasing inion-basion distances, the indices for E. h. hemionus show the contrary tendency (as indicated by the oppositely sloping regression line).

Looking at other species of Equus, in E. africanus the minimum occipital width corresponds to the distance between the external occipital protuberances instead of between the mastoid crests as is the case in E. hemionus. (Among the half-asses measured, only the E. h. onager skull B.M. 1966.3.14.1 has the minimum width a bit higher than between the mastoid crests). The skull of E. africanus has a wider and shorter nuchal tuberosity so that the area between it and the foramen magnum seems larger. To be noted, however, is the fact that in E. hemionus specimen F.M. 26722, the nuchal tuberosity is wide and short as in E. africanus.

The E. przewalskii occipital is both relatively and absolutely wider than those of the half-asses, with a more highly pronounced nuchal tuberosity housed in a deep narrow depression that widens laterally into two symmetrical fossae into which a nutritive foramen opens.

As in E. h. onager (B.M. 39.4784), the basi-occipital of our specimen has a roughly square section and is not rounded as in the other E. hemionus subspecies, in E. africanus, and in E. przewalskii.

Squamous temporal. The articular portion of this element from Shahr-i Sokhta is slightly larger than that of the comparative asiatic wild half-asses, and although morphologically fairly similar, it has 1) a greater depression of the digital impression, 2) a larger retro-articular process, and above all 3) a condyle that is much more highly developed both in width (54 mm as opposed to 41.5-52 mm) and in an antero-posterior direction (17 mm as opposed to 11.5-14.5 mm). The size of the squamosal, and especially of its condyle, indicates that the individual from which

TABLE 2

List of equid specimens used for comparison

	genus: Equus species: hemionus subspecies:	sex	elements available	provenience	museum and number
a.	hemionus	F	skull	Mongolia	F.M. 26717
b.	hemionus	F	skull	Mongolia	F.M. 26718
c.	hemionus	M	skull	unknown	F.M. 26719
d.	hemionus	M	skull	Mongolia	F.M. 26721
e.	hemionus	F	skull	Mongolia	F.M. 26722
f.	hemionus	?	skull	?Nepal	B.M. 1879.11.21-182*
g.	hemionus	M	skull	N. of Ebi Nor, Dzungaria	B.M. 12.5.7.1
j.	kiang	?	skull	Ladakh	B.M. 1879.11.21.194
k.	kiang	F	skull	Kuku Nor, Tibet	B.M. 94.2.8.3
l.	kiang	F	skull	Lhasa, Tibet	B.M. 1905.6.20.1
m.	kiang	M	skull	Sikkim	B.M. 1858.6.24.150*
n.	kiang	?	skull	Ladakh	B.M. 1851.7.16.4
o.	kiang	F	skull	Ladakh	B.M. 1851.7.16.5
u.	kiang	?	post-cranial	?Ladakh	B.M. 976.e
w.	kulan	M	incomplete post-cr.	unknown	B.M. 1971.2210
v.	khur	?	incomplete post-cr.	Kutch	B.M. 1846.1.10.5-705A
p.	khur	M	complete skeleton	Kutch	B.M. 1957.7.18.1
q.	khur	M	skull	Kutch	B.M. 46.592
r.	khur	M	skull	Kutch	B.M. 40358
h.	onager	F	skull	unknown	B.M. 1966.3.14.1
i.	onager	M	skull (pathological)	Iran	B.M. 39.4784
t.	onager	?	post-cranial	Damghan, Iran near Samnan	F.M. 97880
s.	hemippus	?	complete skeleton	Syria	M.C.Z. 6345
	species:				
y.	przewalskii	F	complete skeleton	unknown	M.C.Z.R. 20275-3/4
x.	africanus	M	complete skeleton	E. Abyssinia	M.C.Z.R. 6488

F.M. = Field Museum of Natural History, Chicago
B.M. = British Museum (Natural History), London
M.C.Z. = Museum of Comparative Zoology, Harvard University, Cambridge, Mass.
M.C.Z.R. = Museo Civico di Zoologia di Roma

* doubtful specific attribution, see Groves and Mazák 1967, pp. 348-351.
** identification letters used in Tables 3 and 4.

it came was rather large. While E. przewalskii has a larger squamosal, its condyle is smaller.
Fused onto the Shahr-i Sokhta squamosal is a small portion of the sphenoid. The preserved portions of the basisphenoid and of the pterygoid crest form almost a right angle with the portion of the presphenoid near the hiatus orbitale. This is also the case in the E. hemionus comparative specimens with the exception of an E.h. kiang skull (B.M. 1858.6.24.150) in which, as also in E. africanus, the angle is an obtuse one. In E. przewalskii the passage from one region to the other takes place gradually.

Incisor teeth. The three right, probably lower, incisor teeth all lack the crown and come from the same individual. They are characterized by considerable narrowing of the apical extremity of the root and are very similar to those of the single specimen of E. h. hemippus (MCZ 6345).

Probable upper left premolar. This tooth consists of the grinding surface with unworn denticles.

Probable second upper left molar (Fig. 2H). This is a fragment of the labial portion including the protocone. The protocone is very long (14 mm) and narrow and has two almost symmetrical lobes.

Third right lower molar (Fig. 2I). This tooth probably came from the same animal as the three incisors. This tooth is incomplete, especially in the anterior portion, so that the parastylid and protoconid are almost completely missing from the occlusal face. The protoconid-hypoconid valley is rather shallow--as is the case in half-asses and asses--amounting to about one-third of the width of the tooth in our specimen. This valley is much deeper in E. przewalskii and occasionally reaches the metaconid-metastylid valley. This character is useful in distinguishing asses and half-asses from horses (Stehlin and Graziosi 1935; Groves and Mazák 1967) and from zebras (Turnbull and Reed 1974) even though there can be exceptions as Turnbull and Reed found in the case of the Palegawra Cave E. hemionus (1974, p. 111).

The entoflexid of our specimen is very long as in the half-asses and horses (Turnbull and Reed 1974) and is asymmetrical with a much more highly developed anterior lobe. In the entoflexid, starting from the antero-internal corner of the hypoconid, there runs an enamel fold in a slightly backward direction towards the metastylid.

The metaconid-metastylid valley is shaped like an open "V" as in the half-asses, but unlike the narrow and pointed "V" of the asses (Stehlin and Graziosi 1935) and zebras (Turnbull and Reed 1974) or the "U"-shaped valley in horses (McGrew 1944).

The hypoconid-hypoconulid valley is wide and shallow.

FIGURE 1. *Occipital measurements of Equus hemionus (half-asses). Vertical axis: maximum breadth; horizontal axis: inion-basion length. The symbols designate the following:*
  a. *E.h. hemippus*
  b. *E.h. onager*
  c. *E.h. khur*
  d. *E.h. hemionus*
  e. *E.h. kiang*
  f. *Shahr-i Sokhta specimen*

Scapula (Fig. 2J). This piece is the ventral portion of a left scapula, morphologically very similar to those from the half-asses used for comparison, but larger. The medial face is almost flat and bears shallow vascular furrows which, however, are less pronounced than those of E. h. hemippus (MCZ 6345) and E. h. khur (B.M. 1957.7.18.1--a very old animal). Unlike the comparative specimens, the Shahr-i Sokhta scapula has almost no roughness along the caudal edge for the attachment of the long portion of the triceps brachiale. Although it lacks a small portion of the medial edge, the glenoid cavity has an oval shape, as in the hemiones, with the antero-posterior diameter being the larger. Like the scapula of E. h. kulan (B.M. 1971.2210), the supraglenoidal tubercle is proportionately less developed in height than width when compared to the other half-asses; in E. h. hemippus (MCZ 6345), however, the tubercle is wider than it is high.

The E. przewalskii scapula, unlike the Shahr-i Sokhta specimen, has a medial face that is concave in the area lying between the caracoid apophysis and the articular angle, has more pronounced vascular furrows, an almost circular glenoid cavity, a higher supraglenoidal tubercle, and a less flattened spine. The scapula of E. africanus does not show any remarkable differences from the Shahr-i Sokhta specimen except that it has a higher supraglenoidal tubercle (as in E. przewalskii).

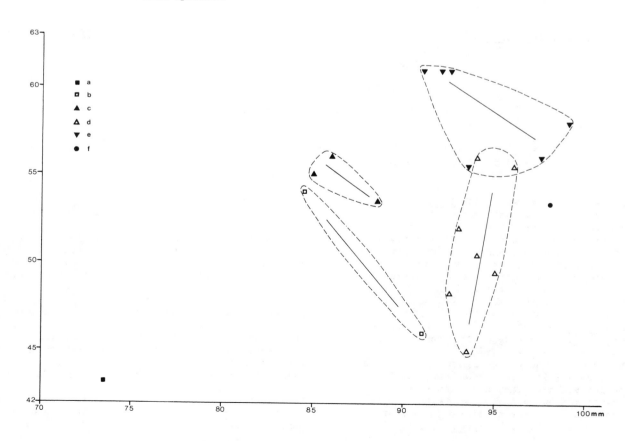

TABLE 3

Skull measurements for Shahr-i Sokhta and modern half-asses (in millimeters)

Occipital	aa**	a	b	c	d	e	f
1. inion-basion distance	(98.0)	93.5	94.0	94.0	95.0	96.0	92.5
2. inion-opisthion distance	(67.0)	56.5	63.5	65.0	59.0	62.0	57.5
3. minimum width	53.4	45.0	50.5	56.0	49.5	55.5	48.3
4. maximum width	(86.0)	89.5	100.0	91.5	94.0	94.0	92.5
5. width of the condyles	69.0	66.0	71.0	70.0	74.0	72.0	67.0
6. width of foramen magnum	31.8	33.0	30.0	31.2	33.3	32.0	30.0
7. height of foramen magnum	29.8	34.5	27.5	28.0	31.8	33.5	35.0
8. index: 3 x 100/1	54.5	48.4	53.7	59.6	52.1	57.8	52.2

	g	h	i	j	k	l	m
1.	93.0	91.0	84.5	92.0	93.5	92.5	91.0
2.	59.5	57.5	56.0	58.0	63.0	59.0	55.0
3.	52.0	46.0	54.0	61.0	55.5	61.0	61.0
4.	96.0	90.5	93.0	95.5	98.0	(105.0)	104.5
5.	77.5	66.0	68.5	71.5	71.5	73.0	78.5
6.	32.2	30.5	30.0	28.5	31.0	35.5	32.0
7.	30.2	30.0	27.5	27.5	26.0	30.5	33.5
8.	55.9	50.5	63.9	66.3	59.4	65.9	67.0

	n	o	p	q	r	s
1.	97.5	99.0	86.0	85.0	88.5	73.5
2.	60.0	65.0	53.5	51.5	56.0	50.0
3.	56.0	58.0	56.0	55.0	53.5	43.2
4.	--	106.5	100.0	105.0	102.5	86.0
5.	76.0	76.0	64.5	68.5	67.8	66.8
6.	31.5	31.0	28.0	32.0	28.5	30.0
7.	31.5	33.0	27.0	29.5	30.5	25.0
8.	57.4	58.6	65.1	64.7	60.5	58.8

Squamous Temporal	bb	a	b	c	d	e	f
1. maximum width	62.0	57.0	57.0	53.5	52.0	59.0	58.0
2. width of condyle	54.0	49.5	44.5	43.5	44.0	46.0	46.5
3. minimum ant.-post. length of condyle	17.0	13.2	12.5	12.5	12.0	12.2	13.7

	g	h	i	j	k	l	m
1.	55.5	56.5	57.0	52.5	57.0	60.0	52.5
2.	46.5	41.5	50.0	42.5	48.0	48.0	44.5
3.	10.5	13.5	11.5	12.5	11.5	14.5	12.0

	n	o	p	q	r	s
1.	57.0	57.0	54.0	60.5	59.0	62.5
2.	52.0	47.5	46.5	51.5	50.5	46.0
3.	12.0	13.5	12.7	12.5	14.5	11.5

** see Table 1 for identification of provenience of Shahr-i Sokhta specimens
see Table 2 for identification of comparative specimens

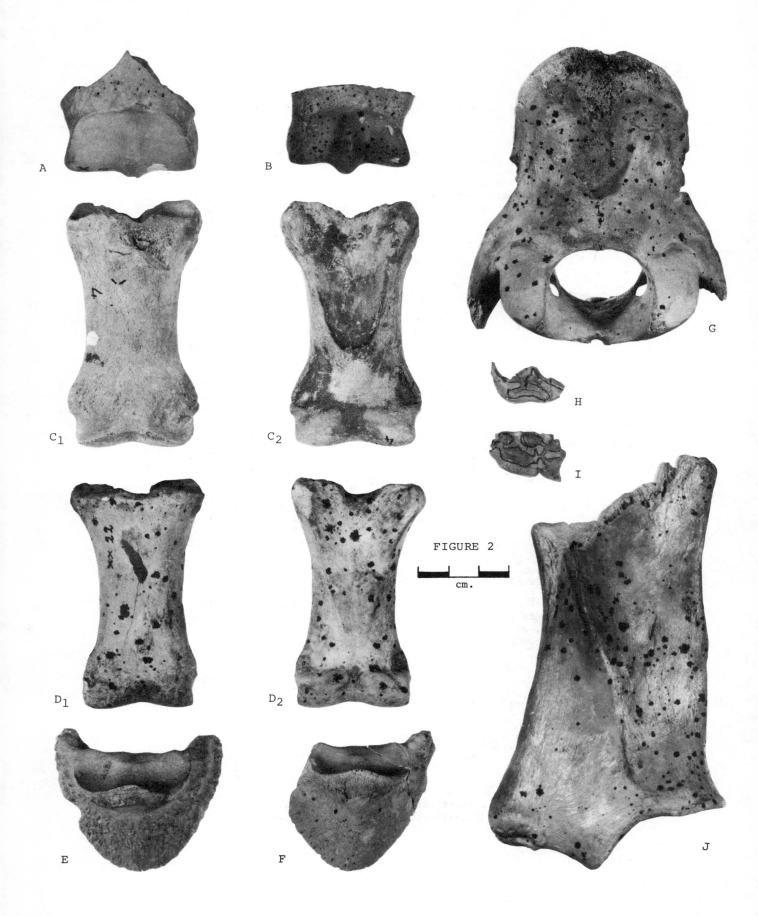

FIGURE 2

FIGURE 2. *Bones of Equus hemionus from Shahr-i Sokhta (scale in centimeters is approximate).*
  A. *left metacarpal [provenience X.8], dorsal view of distal end*
  B. *right metatarsal [XX.26] from juvenile animal, dorsal view of distal epiphysis*
  C. *left posterior phalanx 1 [XII.4]*
    1. *dorsal view*
    2. *plantar view*
  D. *right anterior phalanx 1 [XX.22]*
    1. *dorsal view*
    2. *plantar view*
  E. *posterior phalanx 3 [s.c.(a)], dorsal view*
  F. *posterior phalanx 3 [s.c.(b)], dorsal view*
  G. *occipital [XX.21], caudal view*
  H. *left upper second molar (?) [RDC.2], view of occlusal surface*
  I. *right lower third molar [CX.24], view of occlusal surface*
  J. *left scapula [s.c.], lateral view of distal end*

Metapodials (Fig. 2A-B). These are represented by two distal epiphyses including a left, probably anterior one from an adult individual (provenience X.8) and a right, probably posterior one from a young individual (provenience XX.26). As can be seen in Table 4, our specimens have very large dimensions. In particular, the metacarpal exceeds in size the largest metapodials of the modern forms of E. hemionus examined; the metatarsal, even though from a young animal, is comparable in size to the largest comparative specimen.

Pelvis (Fig. 3D). There are two specimens--a right (provenience XX.23-24) and a left (provenience "s.c."). The right portion lacks part of the neck and all of the wing of the ilium, almost all of the tabula of the ischium, and the caudal branch of the pubis. The left portion consists solely of the acetabular portion of the ilium.

The size of the right pelvic fragment is greater than that of any of the comparative half-asses. Although closely resembling them, the neck of the ilium is larger, its latero-superior edge is not so straight, and the grooves for insertion of the anterior rectus muscle of the thigh are much deeper. In E. h. khur (B.M. 1957.7.18.1) these grooves are almost completely absent. The acetabulum of our specimen is round in shape, as in E. h. hemippus (MCZ 6345) and E. h. khur (B.M. 1957.7.18.1), whereas it is slightly elliptical in the others with the cranio-caudal diameter being the larger.

Femur (Fig. 3A). This right proximal end is considerably larger than those of the comparative specimens of E. hemionus. Although morphologically very similar to the latter, the top of the trochanter of our femur is shorter than in E. h. kulan (B.M. 1971.2210). The trochanter notch is like that of E. h. kulan but narrower than that in E. h. khur (B.M. 1957.7.18.1) and larger than that in the other half-asses.

In femurs of E. africanus and E. przewalskii, the top of the trochanter is higher and narrower, and the intertrochanter notch is less pronounced than in the hemiones. In addition, in E. przewalskii, the angle formed by the upper edge of the third trochanter with the lateral edge of the diaphysis is obtuse and not acute as in E. africanus and the half-asses.

Patella (Fig. 3B). This bone comes from the left leg and is almost identical in size to that of the half-asses except for E. h. hemippus (MCZ 6345) which is much smaller. Of those compared, only the patella of E. h. onager (F.M. 97880) is slightly different from the Shahr-i Sokhta specimen, being relatively and absolutely higher and having a pronounced convexity of the upper edge of the articular surface, a feature much reduced in the others.

The patella of E. przewalskii has a more sharply angled and centrally placed base, with the medial and lateral edges almost equal in length.

Calcaneum (Fig. 3C). This is a right specimen in which almost the whole proximal extremity is missing. In our specimen, as in E. h. hemippus (MCZ 6345), the maximum width, taken between the tuberosity of insertion and the medial edge of the sustentaculum, is equal to the maximum dorsoplantar depth while in the other modern half-ass specimens the width is always greater than the depth.

Compared with E. africanus and E. przewalskii, the dorsal edge of the great apophysis of the Shahr-i Sokhta calcaneum is thinner and, when compared with the horse only, has a smaller articular face. It is significant that in E. h. kiang (B.M. 976.e) this face is separated from that situated on the beak. Furthermore, when compared with E. africanus and E. przewalskii, the articular surface on the sustentaculum of the Shahr-i Sokhta specimen is narrower and higher and contacts more intimately the surface situated under the beak.

Phalanx 1 (Fig. 2C-D). There are two first phalanges from Shahr-i Sokhta, a right anterior one and a left, probably posterior one with a small portion of the dorsal edge of the proximal end missing.

The anterior phalanx resembles those of the various half-ass specimens used for comparison. Whereas its length (81 mm) is close to the average value for the modern material, the index of robustness (minimum width x 100/maximum length) is slightly higher than the greatest value for the comparative specimens (34.8 as opposed to 25.0-34.0). Indeed, the value of this index for the Shahr-i Sokhta phalanx is included in the range of the asses (28.0-37.3) while it is lower than that for the horses (35.3-49.4). The phalanges of the true asses, however, are shorter than our specimen. [For dimensions and index of robustness of the first phalanges of half-ass, ass (E. asinus, africanus, somaliensis, and palestinae) and horse (E. caballus var. and przewalskii), we have used, in addition to the specimens noted in Table 2, values quoted in Gromova 1949; Boessneck and von den Driesch 1967; Ducos 1968; 1970; Kolb 1972; Boessneck and Krauss 1973; Clutton-Brock 1974, Krauss 1975.]

Our posterior phalanx differs morphologically

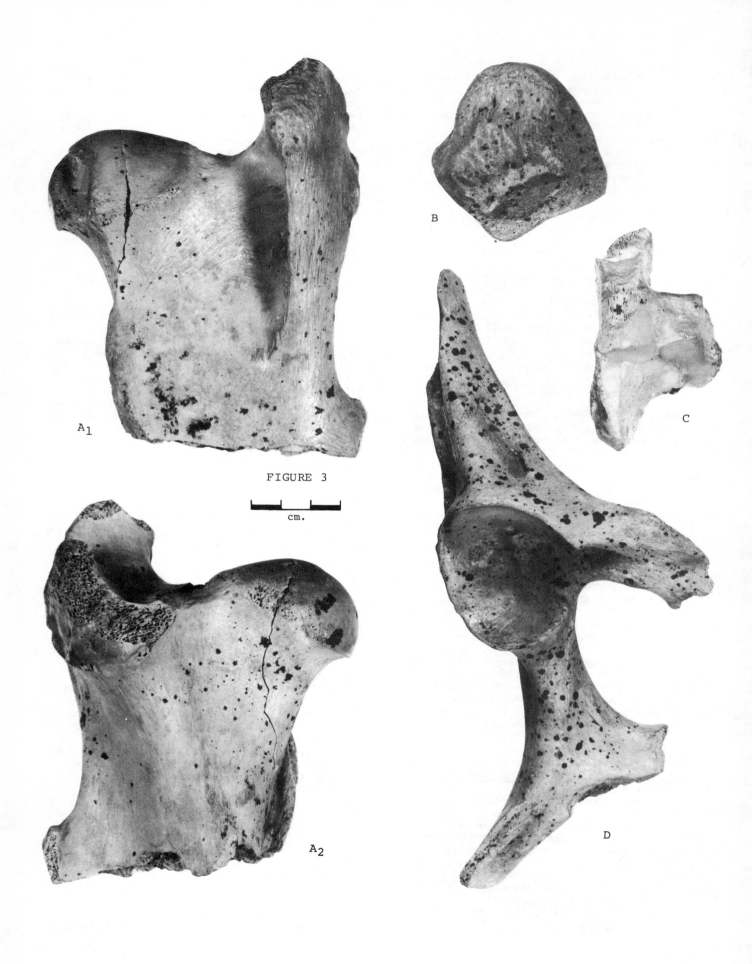

FIGURE 3

FIGURE 3. *Bones of Equus hemionus from Shahr-i Sokhta (scale in centimeters is approximate).*
- A. *right femur [provenience 's.c.']*
  1. *caudal view of proximal end*
  2. *cranial view of proximal end*
- B. *right patella [s.c.], dorsal view*
- C. *right calcaneum [s.c.], dorsal view*
- D. *right pelvis [XX.23-24], lateral view*

from those of the comparative specimens of half-ass only in that it has a much deeper ligament insertion scar on the palmar surface. Because of its length (87 mm), it must have belonged to a large individual. The only posterior phalanx of similar size published in the literature is from an animal found near Liaptichev in the north Caucasus (86.5 mm, Gromova 1949). The index of robustness of our specimen is slightly higher than the maximum value found in the half-asses (35.6 as opposed to 30.7-34.9) but lower than that for horses (36.6-51.3) with one exception (33.9, Kolb 1972, although the value for the minimum width of the shaft reported there may be a misprint). The index for our posterior phalanx, but certainly not its size, is included in the range for the true asses (31.0-38.1).

In the scatter diagram, Figure 4, the values for maximum length and minimum width of the first phalanges of half-asses, asses, and horses are presented along with those for the Shahr-i Sokhta specimens (values taken from Table 4 and the various references noted above). Values for the horses lie in the upper part of the diagram with those for *E. przewalskii* being close to the lower margin of the horse cluster. Values for the asses and half-asses overlap partially and lie in the lower part of the diagram, clearly separated from those for horse by a broad band in which fall the dimensions for the Shahr-i Sokhta specimens and for two other probable half-ass phalanges from Liaptichev and Altai Cave (Gromova 1949). [Note that Gromova takes the width measurement at a point one-half way along the bone; this measurement, however, should provide a value nearly identical to minimum width.]

Taking into consideration the index of robustness of the anterior first phalanx (as opposed to the raw dimensions), it is evident that there is a great overlap between the values of ass and half-ass and a small overlap between those of ass and horse with the values for half-ass being quite distinct from those for horse. The same is true for the posterior phalanges. If one takes the anterior and posterior phalanges together, only the posterior phalanx from Shahr-i Sokhta falls in the range for horse.

FIGURE 4. *Measurements of first phalanges of onager, donkey, and horse. Vertical axis: minimum breadth; horizontal axis: maximum length. The symbols designate the following:*
- a. *Equus caballus (various forms)*
- b. *E. przewalskii*
- c. *E. hemionus (various subspecies)*
- d. *E. asinus (various subspecies)*
- e. *E. palestinae*
- f. *E. hemionus from Shahr-i Sokhta*
- g. *field of variation for E. asinus*
- h. *field of variation for E. przewalskii*

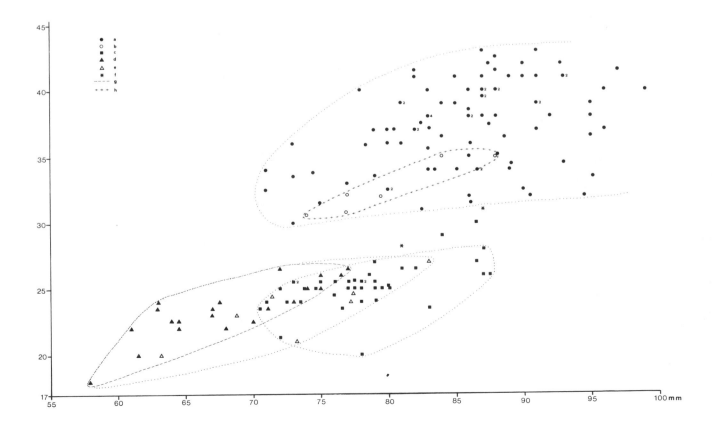

TABLE 4

Post-cranial measurements for Shahr-i Sokhta and modern
half-asses (in millimeters)

Scapula	cc**	t	u	v	w	s	x	y
length ventral angle	79.5	76.0	75.5	74.0	74.5	64.0	69.5	86.3
length glenoid cavity	52.0	46.0	48.5	45.0	45.5	39.5	45.0	51.5
width glenoid cavity	(42.2)	40.0	39.5	39.0	39.0	37.0	39.2	53.1
minimum length neck	53.2	50.0	52.0	48.0	49.0	44.0	51.0	56.7
width supraglenoid tubercle *	28.2	26.3	25.0	26.5	27.0	22.0	26.5	28.5
height supraglenoid tubercle *	29.5	33.5	32.5	34.5	27.5	20.0	36.0	35.0

Pelvis	dd	t	u	v	p	w	s	x	y
length acetabulum	53.3	50.0	52.0	48.0	49.0	51.5	44.5	51.0	61.2
width acetabulum	53.0	47.0	48.0	45.0	49.5	50.5	43.0	48.0	57.0
min. width ischium	21.1	18.0	17.0	19.5	22.2	18.5	17.5	18.0	25.0

Femur	ee	t	u	p	w	s	x	y
max. proximal width	107.0	98.0	98.5	96.5	97.5	87.0	92.0	108.5
ant.-post. diameter articular head	50.0	46.0	48.0	45.0	46.0	40.0	45.3	55.0

Patella	ff	t	u	p	w	s	x	y
height, basis-apex	62.0	64.2	60.5	61.5	62.0	52.2	55.0	63.3
width	60.8	57.8	57.2	60.5	59.5	49.5	55.6	62.6
ant.-post. depth	28.7	27.3	27.5	25.5	28.0	21.0	26.0	32.5
width artic. surf.	46.2	49.3	42.5	44.0	42.5	39.5	41.5	48.9

Calcaneum	gg	t	u	p	s	x	y
height at beak	46.5	41.5	43.5	42.2	39.0	41.0	47.0
max. width of distal end	46.5	43.0	46.0	48.0	39.0	44.5	51.5

Metacarpal	hh	t	u	p	w	s	y
width dist. artic. surf.	43.2	39.0	38.0	41.0	40.2	36.2	44.5
max. ant.-post. diameter dist. artic. surface	33.2	29.0	30.0	27.8	30.0	28.0	33.0

Metatarsal	ii	t	u	p	w	s	x	y
width dist. artic. surf.	40.0	37.5	38.2	37.8	38.0	34.5	38.5	44.2
max. ant.-post. diameter dist. artic. surface	30.7	30.0	31.5	29.0	30.5	28.3	29.2	34.8

TABLE 4

(continued)

Anterior Phalanx 1	jj	t	u	p	w	s	x	y
1. max. length	81.0	77.5	87.5	75.0	79.0	72.0	75.0	77.0
2. min. width shaft	28.2	25.6	26.0	25.5	25.5	21.3	26.0	32.1
min. depth	17.0	17.0	17.5	17.2	17.2	15.0	17.5	19.7
max. prox. width	44.7	42.1	41.8	44.5	41.5	36.0	40.0	49.3
prox. depth	31.8	31.0	32.9	30.2	30.7	27.5	29.5	34.0
max. distal width	39.4	37.8	37.9	39.0	38.0	32.2	35.0	43.1
distal depth	21.5	20.0	22.2	19.8	20.5	18.2	20.0	23.0
index (2 x 100/1)	34.8	33.0	29.7	34.0	32.3	30.0	34.7	41.7

Posterior Phalanx 1	kk	t	u	p	w	x	y
1. max. length	87.0	71.0	80.0	70.5	73.0	75.0	76.8
2. min. width shaft	31.0	24.0	25.2	23.5	25.5	25.0	30.8
min. depth	20.8	15.6	17.5	16.5	17.0	18.0	18.0
max. prox. width	45.6	42.8	42.0	43.8	43.0	42.0	50.9
prox. depth	--	31.2	33.5	32.0	31.0	29.8	36.6
max. distal width	44.5	36.2	36.0	37.8	36.5	34.0	42.0
distal depth	24.2	19.0	21.8	19.2	19.8	19.5	22.5
index (2 x 100/1)	35.6	33.8	31.5	33.3	34.9	33.3	40.1

Posterior Phalanx 3	ll	mm	u	p	s	x	y
max. length	50.0	(47.0)	53.0	47.0	40.0	40.5	60.0
max. width	57.5	(52.5)	58.0	52.5	42.3	48.5	68.8
max. height	37.5	36.0	35.0	32.5	29.0	30.0	38.0
width artic. surf.	39.0	34.2	34.5	37.5	31.5	32.0	42.2

Posterior Distal Sesamoid	nn	u	s	x	y
length	11.0	15.0	10.0	12.0	13.7
width	37.9	38.5	29.5	36.5	41.6
height	10.7	11.2	9.2	10.0	11.2

* coracoid apophysis included

** see Table 1 for identification of provenience of Shahr-i Sokhta specimens
   see Table 2 for identification of comparative specimen

It seems clear from these distributions that it should be possible to distinguish the phalanges of ass and half-ass from those of horse by considering together the dimensions and indices. Even if there is overlap between horse and ass in the index, the two can be separated by using the dimensions. Distinguishing between the phalanges of E. hemionus and the asses is more difficult unless the dimensions and indices lie within the exclusive range of one or the other because generally the half-ass phalanges are larger and have a smaller index than those of the asses. The same is true for the metapodia (Hilzheimer 1941) where it is possible to separate horses from asses and half-asses using the index of robustness but more difficult to separate asses and half-asses using the same means.

Phalanx 3 (Fig. 2E-F). Two left posterior hoof cores were recovered, the smaller ("s.c."b) lacking the internal retroassalis apophysis. It has been possible to compare these specimens with examples from three half-asses (E. h. khur B.M. 1957.7.18.1, E. h. kiang B.M. 976.e, and E. h. hemippus MCZ 6345). They were found to be intermediate in size between E. h. khur and E. h. kiang. The margin of the soles of both specimens is fairly pointed and lacks a median notch. In E. h. hemippus, the margin is rounder. In our specimens, the pyramidal process is poorly developed in relation to the dorsal face, less indeed than in E. h. kiang, so that, in profile, the dorsal edge dominates the process itself as in the E. hemionus specimen illustrated by Stehlin and Graziosi (1935, Table 3, Fig. 12). From this last, our phalanges differ only in that they are a little wider and have more pronounced dorsal furrows. In E. h. khur, the pyramidal process is situated further forward and has less of an inclined articular surface.

The hoof core of E. przewalskii has a much rounder sole edge and a much less concave sole surface.

Distal sesamoid. The only specimen found, a right posterior one, closely resembles those of E. h. hemippus (MCZ 6345) and E. h. kiang (B.M. 976.e) and, like them (unlike that of E. przewalskii), its lateral extremities are turned in a proximal direction.

General considerations

Comparison of the Shahr-i Sokhta equid material with skeletal material of half-asses from other archaeological sites is possible only to a limited degree given the incompleteness of our specimens and the scarcity of morphological descriptions and metrical data in the published literature. Too often authors are more interested in reporting the frequency of a species and its change in abundance through time than in recording characteristics of the individual specimens. Information about living populations of E. hemionus is little better, however, these rare animals having been scarcely studied either osteologically or osteometrically.

Half-asses appear in the Upper Pleistocene layers of some asiatic sites. Gromova (1949) reports the presence of large-sized E. hemionus at Altai Cave and Afontova Gora, these being only slightly shorter than those from Shahr-i Sokhta. She also reports smaller half-asses from Ordos in inner China.

The E. hemionus remains from Palegawra Cave (northwestern Iraq, 12,000 B.P.) considered by Turnbull and Reed (1974) to be larger than the extant Persian onager ("ghor-khar"), are, in fact, smaller than those from Shahr-i Sokhta. Comparing dimensions, it is evident that our metacarpus and posterior first phalanx are larger than the largest specimens from Palegawra while our juvenile metatarsus and anterior first phalanx are nearly the size of the largest pieces from the cave.

Among the later sites, Anau in southern Turkmenia (3000-1500 B.C.) is particularly important (Duerst 1908; Gromova 1949). The limb bones of half-ass found there are, in general, smaller than those from Shahr-i Sokhta with only the greatest values for the anterior first phalanx and for the posterior third phalanx being slightly greater than those for our equid. Ermolova (1970) gives measurements for some metatarsals from other south Turkmenia sites (Šor-depe, Altyn-depe) of the late third millennium B.C. and they are almost as large as those of our juvenile specimen.

From Bastam in northwest Azerbaijan (2200-700 B.C.) come some half-ass bones among which the anterior first phalanx is larger, the posterior ones smaller, and the others about the same size as those from Shahr-i Sokhta (Boessneck and Krauss 1973; Krauss 1975). From the same area, Kolb (1972) reports some measurements of onager bones collected in deposits of the eighth to tenth century A.D. at Takht-i Suleiman. The metatarsus and two posterior phalanges from that site are smaller than the corresponding bones from Shahr-i Sokhta.

Gromova (1949) presents information also about probable half-asses found in archaeological sites in the Ukraine and the Caucasus. Near Odessa, in deposits of 3000-2000 B.C., an anterior first phalanx smaller than ours was found, and near Kiev, a metatarsus dating to the eleventh and twelfth century A.D. was recovered which is slightly larger than our juvenile specimen. A metatarsus and two posterior first phalanges have been found near Liaptichev in Bronze Age sediments (1500-1000 B.C.). Of these, one very large phalanx (86.5 mm in length) is almost the same size as its counterpart at Shahr-i Sokhta, while the other may have come from a female animal and is thus not so large. Finally, from the second to third century A.D. levels at Isti-Sou comes an anterior phalanx which approximates in size that from Shahr-i Sokhta.

No data are available on the size of the E. hemionus bones from Khuzistan (southwest Iran: Hole, Flannery, and Neely 1969), from some Turkmenia sites (Calkin 1970; Ermolova 1968, 1970, 1972), from the Quetta Valley (Fairservis 1956), or from Umm Dabaghiyah (north Iraq: Bökönyi 1973) since no measurements are reported.

As far as comparison with living subspecies of half-ass are concerned, data reported in Gromova (1949) and study of collections in various museums have permitted the author to conclude that the size of the Shahr-i Sokhta specimens is close to the size of comparable bones from the largest living hemiones and, for some bones, even larger (squamous temporal, scapula, femur, metacarpus, and posterior first phalanx). The E. hemionus living in Sistan between the end of the fourth and the beginning of the third millennium B.C. was thus an animal of considerable size.

At present the half-ass has disappeared from Sistan (Firouz 1974) so its morphological and metrical characteristics are known only through the finds from Shahr-i Sokhta. Nevertheless, it may be supposed that in a region like Sistan, with its peculiar geomorphic features and climatic conditions, a population of half-asses, isolated from the original stock, could have developed particular characteristics. Indeed an isolated environment like Sistan seems favorable for microevolutionary processes as is also suggested by the regional subspecies defined for modern wild mammals (otter, gazelle, etc., Lydekker 1910).

References

Azzaroli, A.
　1966　"Pleistocene and living horses of the Old World. An essay of classification based on skull characters," *Palaeontographia Italica,* vol. 66 (n.s. 31), pp. 1-15.

Bökönyi, S.
　1972　"Once more on the osteological differences of the horse, half-ass, and the ass," in L. Firouz, *The Caspian Miniature Horse of Iran,* pp. 12-23. Miami, Field Research Projects.
　1973　"The fauna of Umm Dabaghiyah: a preliminary report," *Iraq,* vol. 35, pp. 9-11.

Boessneck, J.
　1973　"Tierknochenfunde vom Zendan-i Suleiman (7.Jahrhundert v. Christus)," *Archaeologische Mitteilungen aus Iran,* n.F., vol. 6, pp. 95-111.

Boessneck, J. and A. von den Driesch
　1967　"Die Tierknochenfunde des frankischen Reihengräberfeldes in Kleinlangheim, Landkreis Kitzingen," *Zeitschrift für Säugetierkunde,* vol. 32, no. 4, pp. 193-215.

Boessneck, J. and R. Krauss
　1973　"Die Tierwelt um Bastam/Nordwest-Azerbaidjan," *Archaeologische Mitteilungen aus Iran,* n.F., vol. 6, pp. 113-133. Berlin.

Bourdelle, E.
　1955　"Perissodattiles," in Grasse, editor, *Traité de Zoologie,* vol. 17, no. 1, pp. 1002-1126.

Calkin, V.I.
　1970　"Drevnejšie domašnie životnye Srednej Azii. Soobščenie I," *Bjulleten' Moskovskogo Obščestva Ispytatelej Prirody - Otdel Biologičeskij,* vol. 75, no. 1, pp. 145-159.

Clutton-Brock, J.
　1974　"The Buhen horse," *Journal of Archaeological Science,* vol. 1, no. 1, pp. 89-100.

Ducos, P.
　1968　*L'Origine des Animaux Domestiques en Palestine.* Publications de l'Institut de Préhistoire de l'Université de Bordeaux, Mémoire no. 6. Bordeaux.
　1970　"The Oriental Institute excavations at Mureybit, Syria: preliminary report on the 1965 campaign. Part IV: les restes d'equidés," *Journal of Near Eastern Studies,* vol. 29, pp. 273-289.

Duerst, J.U.
　1908　"Animal remains from the excavation at Anau and the horse of Anau in its relation to the history and to the races of domesticated horses," in R. Pumpelly, editor, *Explorations in Turkestan, Expedition of 1904,* pp. 341-442. Washington.

Ermolova, N.M.
　1968　"Kostnye ostatki mlekopitajuščih iz poselenij epohi eneolita i bronzy Južnogo Turkmenistana," *Karakumskie Drevnosti,* vol. 1, pp. 48-53.
　1970　"Novye materialy po izučeniju ostatkov mlekopitajuščih iz drevnih poselenij Turkmenii," *Karakumskie Drevnosti,* vol. 3, pp. 48-53.
　1972　"Ostatki mlekopitajuščih iz drevnih pamjatnikov južnoj Turkmenii po raskopkam 1970 goda," *Karakumskie Drevnosti,* vol. 4, pp. 177-182.

Fairservis, W.A.
　1956　"Fauna and Flora," Appendix 4 in *Excavations in the Quetta Valley, West Pakistan.* Anthropological Papers of the American Museum of Natural History, vol. 45, part 2, pp. 382-383.

Firouz, E.
　1974　*Environment Iran.* Tehran.

Gromova, V.
　1949　*L'Histoire des Chevaux (genre Equus) de l'Ancien Monde. Première partie: Revue et description des formes.* Akademiia Nauk SSSR Paleontologičeskii Institut, Trudy, vol. 42, pp. 1-373 (in Russian). [French translation: Annales du Centre d'Etudes et de Documentation Paléontologiques, no. 13. Paris, 1955.]

Groves, C.P.
　1974　*Horses, Asses, and Zebras in the Wild.* London.

Groves, C.P. and V. Mazák
　1967　"On some taxonomic problems of Asiatic wild asses; with the description of a new subspecies (Perissodactyla; Equides)," *Zeitschrift für Säugetierkunde,* vol. 32, no. 6, pp. 321-355.

Grzimek, B. (editor)
　1973　*Vita degli animali,* vol. 12 (Mammiferi 3). Varese.

Hilzheimer, M.
　1941　*Animal Remains from Tell Asmar.* Studies in Ancient Oriental Civilization, no. 20. Chicago.

Hole, F., K.V. Flannery, and J.A. Neely
1969 *Prehistory and human ecology of the Deh Luran Plain. An Early Village Sequence from Khuzistan, Iran.* Memoirs of the Museum of Anthropology, University of Michigan, vol. 1. Ann Arbor.

Kolb, R.
1972 "Die Tierknochenfunde vom Takht-i Suleiman in der iranischen Provinz Aserbeidschan (Fundmaterial der Grabung 1969)." Dissertation. München.

Krauss, R.
1975 "Tierknochenfunde aus Bastam in Nordwest-Azerbaidjan/Iran (Fundmaterial der Grabungen 1970 und 1972)." Dissertation. München.

Lydekker, R.
1910 "The gazelles of Seistan," *Nature,* pp. 201-202.

McGrew, O.P.
1944 "An early Pleistocene (Blancan) fauna from Nebraska," *Field Museum of Natural History Geological Serial,* vol. 9, pp. 33-69.

Misonne, X.
1959 *Analyse Zoogéographique des Mammifères de l'Iran.* Mémoires de l'Institut Royal des Sciences Naturelles de Belgique, deuxième série, fasc. 59. Bruxelles.

Stehlin, H.G. and P. Graziosi
1935 *Ricerche sugli Asinidi fossili d'Europa.* Mémoires de la Societé Paléontologique Suisse, vol. 56. Bâle.

Turnbull, P.F. and C.A. Reed
1974 *The fauna from the terminal Pleistocene of Palegawra Cave, a Zarzian occupation site in northeastern Iraq.* Fieldiana: Anthropology, vol. 63, no. 3, pp. 81-146. Chicago.

Zeuner, F.E.
1963 *A History of Domesticated Animals.* London.

THE BONE REMAINS OF GAZELLA SUBGUTTUROSA FROM SHAHR-I SOKHTA

Bruno Compagnoni, Servizio Geologico d'Italia
Rome

Gazella (Trachelocele) subgutturosa Guldenstaedt, 1780

Comparison of the gazelle remains from Shahr-i Sokhta with modern specimens (Table 1) shows that our material resembles most closely that classified as Gazella subgutturosa. The slight morphological differences that exist will be dealt with below, but these lie well within the bounds of intraspecific variability.

The Remains from Shahr-i Sokhta

Thirty-eight gazelle bones, all very well preserved, have been identified from Shahr-i Sokhta, all from adult individuals (Table 2). A few of these bones bear traces of butchering and burning. In addition, at least five Buff Ware sherds bearing representations possibly of gazelle have been found (Fig. 3). The bones of gazelle from Shahr-i Sokhta are described below and their measurements appear in Tables 3 and 4.

Neurocranium (Fig. 1A). This important find, consisting of the portion from the frontal bone through the occipital and one intact and one fragmentary horn core, comes from on top of the artificial fill in Room LXXXVI. This fill includes Phase 8 pottery and small finds probably deposited there at the end of Period I (about 2800 B.C.) when the whole area to the north of Street I in the Eastern Residential Area was made into an empty space between two buildings. The cranium should thus date to the end of Phase 8.

Comparison of the Shahr-i Sokhta neurocranium with skulls from G. subgutturosa (B-2-A/MO/20) and G. dorcas (B-2-B/KA/3), both from Iran, clearly points to identification with the former species. The horn cores on our neurocranium also appear identical to those found isolated, and all appear much more laterally divergent than is the case for G. dorcas. It should be noted, however, that according to B. O'Reagan (Department of Environmental Conservation, Tehran: personal communication) the distance between the horn cores at their base in our specimen falls outside of the range for modern G. subgutturosa and within that for G. dorcas; we consider this feature to be explainable, however, as intraspecific variation.

In the occipital region of our specimen, and of G. subgutturosa, the interparietal is less developed craniocaudally and forms with the exoccipital a wider angle than is customary in G. dorcas, thus displacing the condyles backwards. According to B. O'Reagan (personal communication), this difference is probably due to the fact that the auditory bullae are larger in G. dorcas—the more desert-adapted animal—than in G. subgutturosa. From a caudal view our cranium appears compressed dorso-basally and thus widened at the temporal bones as is the case in G. subgutturosa (just behind the ear orifices). The index of the basioccipital/parietal distance to the width at the temporals is, for the Shahr-i Sokhta find: 81.6, for the G. subgutturosa specimen: 75.3, and for the G. dorcas skull: 93.5 (Tehran specimen measurements).

Horn Cores (Fig. 1B and 2I). These are the most numerous gazelle remains from Shahr-i Sokhta. A total of eleven incomplete specimens have been found: five right and six left; two lack only a small portion of the apex.

Morphologically all the horn cores are very similar. They curve regularly in a posterior direction, and where part of the frontal and related sagittal suture are still present, it can be seen how the horns diverge laterally to a considerable degree with respect to the sagittal plane. In the more complete specimens there is a slight twisting in a heteronymous sense, with the twisting being most pronounced in the specimen from provenience XIG/H.4-6 LXX. The cores are all compressed medio-laterally with the lateral surface being almost flat and medial surface slightly convex. Numerous deep longitudinal furrows run along almost the entire length and generally on all sides of the bones surface although in a few specimens such grooves are missing from the medial face. One or two of the wider and deeper furrows can be seen posteriorly.

As far as size is concerned, our horn-core dimensions lie within the range of variability defined by the specimens from Tell Asmar (Hilzheimer

TABLE 1

List of gazelle specimens used for comparison (all male)

*	genus: Gazella species:	elements available	provenience	museum and number
tt.	subgutturosa	complete	Iran	Iran B-2-A/MO/20
uu.	subgutturosa	incomplete post-cr.	Altai	B.M. 90.4.20.11-1702g
vv.	subgutturosa	incomplete post-cr.	???	B.M. (no number)
ww.	bennetti	skull	Kutch	B.M.1935.12.21.10-496
xx.	bennetti	skull	Palanieny(?)	B.M.24.10.5-35
yy.	dorcas	complete	Iran	Iran B-2-B/KA/3
zz.	dorcas	complete	???	M.C.Z.R. 20282

B.M. = British Museum (Natural History), London
Iran = Department of Environmental Conservation, Tehran
M.C.Z.R. = Museo Civico di Zoologia di Roma

* identification letters used in Table 4.

1941, tab. V), Anau (Duerst 1908, p. 382), Ali Kosh (Hole et al. 1969, Fig. 123a) and the Berlin Museum (Hilzheimer 1941, tab. V). Compared with drawings of the Tell Asmar specimens, our horn cores have a less pronounced curvature, a more acutely angled apex, and a smaller degree of twisting.

The horn cores of G. bennetti are straighter, more regularly conical, and less compressed laterally than those from Shahr-i Sokhta even though part of the lateral surface is almost flat. The longitudinal furrows are also less numerous (one or two) and are situated ventrally.

The horn cores of G. dorcas (M.C.Z.R. 20282) are perceptibly shorter and straighter than those from Shahr-i Sokhta and have no large longitudinal furrows on the surface except in the central zone where there is one for each horn core. The apical portion of the core is also a little more compressed laterally.

Atlas (Fig. 1C). Two different-sized specimens have been found each of which lacks part of the alar expansion of the transverse apophyses. It was possible to compare our pieces only with an atlas of one male G. subgutturosa (B.M. 90.4.20.11.1702g). In size and robustness our larger specimen (from provenience CCXIII.a.10) resembles the British Museum specimen very closely. The difference in size between our two atlases is probably due to sexual dimorphism.

Small morphological differences were noted between our two specimens as well as between those and the comparative specimen particularly with regard to tubercles and articular surfaces. The dorsal tubercle, while very pronounced and roughly triangular in shape in both large specimens, is barely perceptible in the smaller one. The ventral tubercle in this latter specimen is rather elongated craniocaudally and very much flattened laterally. In our larger atlas, this tubercle is swollen distally and narrows at the base, while in the comparative specimen the tubercle is shorter and does not narrow.

In our smaller atlas, the caudal tubercles of the alar expansions of the transverse apophyses are much smaller than in the others. The medial edges of these tubercles extend toward the ventral tubercle cutting off two triangular areas lateral to it which, in both of the large specimens, are greatly depressed. However, whereas in our large atlas the areas are much longer than they are wide, in the comparative specimen the reverse is the case. The articular surface for the occipital in our smaller specimen is not so large as in the other, both in its actual dimensions and because of the greater size of the vertebral foramen which is only slightly smaller than that of the comparative specimen. The articular surfaces for the axis in the larger atlases

FIGURE 1. Bones of *Gazella subgutturosa* from Shahr-i Sokhta (scale in centimeters is approximate).
   A. cranium [provenience XDV.5.NORD]
      1. left side view
      2. anterior view
   B. right horn core [CLXII (6)], medial view (see also Fig. 2I)
   C. atlas [CCXIII.a.10]
      1. caudal view
      2. dorsal view
      3. ventral view

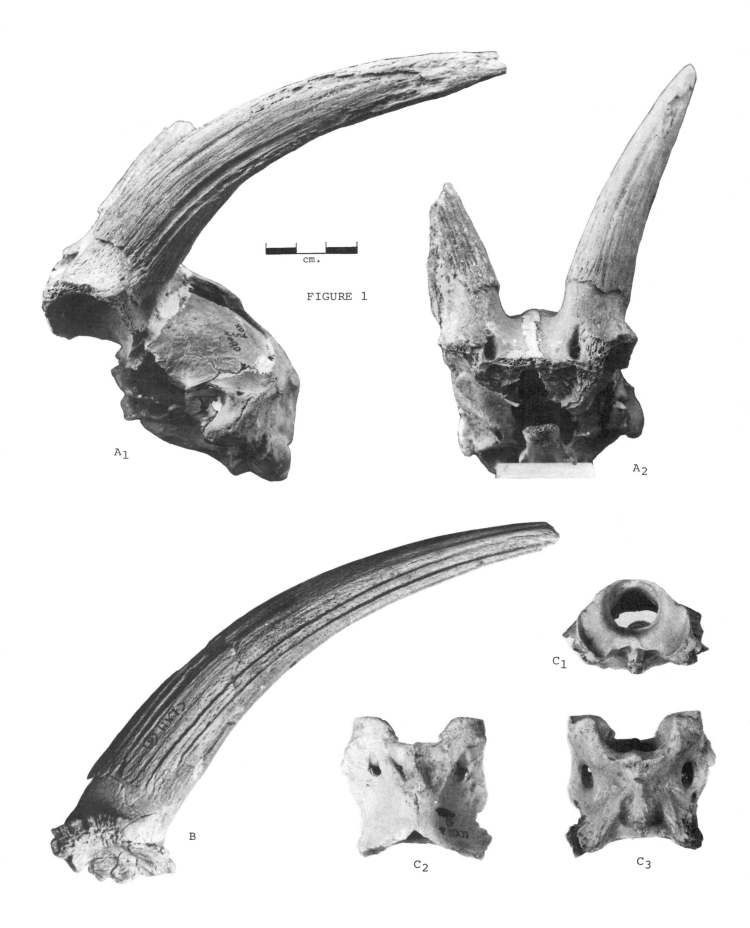
FIGURE 1

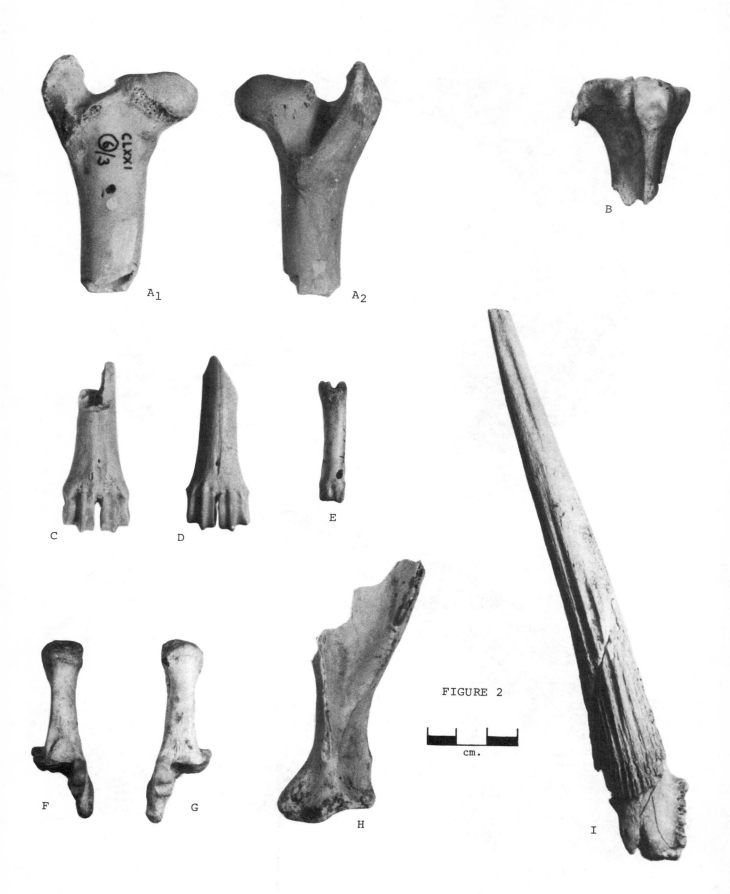

FIGURE 2

FIGURE 2. Bones of *Gazella subgutturosa* from
        Shahr-i Sokhta (scale in centimeters
        is approximate).
   A. *right femur [provenience CLXXI(6)/3]*
      1. cranial view of proximal end
      2. caudal view of proximal end
   B. *right tibia [XX.25], cranial view
        of proximal end*
   C. *left metacarpal [XIH.2(25)], dorsal
        view of distal end*
   D. *left metatarsal [s.c.], dorsal view
        of distal end*
   E. *phalanx 1 [CXX.4], dorsal view*
   F. *left calcaneum [s.c.], dorsal view*
   G. *right calcaneum [CXXIII (2/3)],
        dorsal view*
   H. *left scapula [s.c.] lateral view of
        distal end*
   I. *right horn core [CLXII (6)], anter-
        ior view (see also Fig. 1B)*

touch ventrally over a short stretch (shorter in the comparative specimen) while they are distinctly separate in our smaller specimen.

Scapula (Fig. 2H). We have recovered three scapulae--two right and one left, all missing most of the blade and all roughly equal in size to the larger of the two comparative specimens from the British Museum (90.4.20.11.1702g). The specimens have subcircular glenoid cavities which taper slightly on the lateral side. In our left specimen, the glenoid notch is shallower than in the others. While the supraglenoidal tubercle and the coracoid apophysis are quite similar in our specimens and in the comparative material, the subglenoidal tubercles have slightly different shapes and sizes. In our right specimen ("s.c."), the spine of the scapula is located closer to the caudal edge than in any of the others.

Humerus. We have identified only one distal end of a right humerus which is very close in size to the comparative specimen and to the one illustrated by Gromova (1950). The diaphysis of our specimen is compressed latero-medially and has a small nutritive foramen. This thinness of the diaphysis in the vicinity of the coronoid fossa causes the epicondylic crest to project to a greater extent laterally and to be slightly thicker in the central part when compared with the modern material (including the specimen in Gromova).

Radius. Two radii have been found--a complete right one and a slightly smaller left one consisting of the distal portion only. The complete specimen is the same length as the one illustrated by Gromova (1950) and as the smaller of the two British Museum specimens. Both Shahr-i Sokhta finds still have joined to them part of the ulnar diaphysis and our right radius also has the styloid apophysis firmly joined to it.

Ulna. One left ulna has been found with part of the corpus and the distal end missing. Its morphology has been altered by rodent gnawing particularly in the vicinity of the tip of the olecranon and along the palmar edge of the corpus. Our ulna is smaller than the comparative specimens and appears to be more compressed latero-medially and to have a medial face that is concave instead of flat.

Metapodials (Fig. 2C-D). Four distal ends of metacarpals (one right and three left) and three distal ends of metatarsals (two left and one right) have been found. They are very similar in size and morphology to the comparative specimens.

Pelvis. Two very incomplete pelvic fragments have been identified--an ilio-ischiatic portion and an ischiatic portion. There are no significant differences in size or morphology between them and the comparative specimens. In the specimen from provenience XIC.4, the ischial tuberosity for the insertion of the femoral biceps is much larger than in the other.

Femur (Fig. 2A). The proximal end of one femur was found, practically equal in size and identical in morphology to the larger British Museum specimen and to those illustrated in Gromova (1950) and in Boessneck and Krauss (1973).

Tibia (Fig. 2B). Two right proximal ends of tibia are intermediate in size between the two comparative specimens, but have larger extensor sulci. The specimen from provenience XX.25 has a tibial tuberosity which is bent slightly more laterally than in the comparative material.

Calcaneum (Fig. 2F-G). Four complete specimens of calcanea were found--two right and two left--one of which (from XIB.1) was broken during excavation and is missing the plantar portion. All are roughly the same size and intermediate between the comparative specimens. Compared with the larger British Museum specimen (90.4.20.11.1702g), our specimens have larger articular surfaces where the calcaneum meets the lateral side of the astragalus and the cuboid. In our specimen from provenience CXXIII.2/3 in particular, the articular surfaces are narrow where they join the anterior face of the astragalus and the cuboid.

First Phalanx (Fig. 2E). One first phalanx has been found, intermediate in size between those from the two comparative specimens but morphologically very similar to them. The posterior phalanges illustrated in Boessneck and Krauss (1973) seem to be somewhat wider than our specimen.

Conclusions

On the basis of the morphological descriptions and measurements presented here (Tables 3 and 4), we can confidently assign our gazelle remains to the species *Gazella subgutturosa*. As far as identification to the subspecific level is concerned, Lydekker (1910, p. 202) reports for the "seistanica" form of the Hilmand valley a withers height of "29 inches" as compared with a height of "24-26 inches" for the "subgutturosa" form. The size of the remains from Shahr-i Sokhta suggests that the ancient gazelle was similar in size to the "subgutturosa" form. Recent taxonomic work (Lange 1972), however, places the goitered gazelle of Sistan not even in a separate subspecies but considers it as a form of *G.s. subgutturosa*.

Some students (Misonne 1959; Lay 1967, Firouz 1974) mention the presence of another form of gazelle in central, eastern, and southern Iran. Whether this form is assignable to the species *G. gazella* or *G. dorcas* is undergoing discussion in the literature. At the present stage of

TABLE 2

Bone remains of *Gazella subgutturosa* from Shahr-i Sokhta

(2)	Element:	# of specimens rt.	lft.	Provenience	Period	Phase	Notes[1] a	b	c	d
a.	neurocranium	1		XDV.5 nord	I	8	–	–	–	–
b.	horn core	1		CLXII 6	II	6	–	–	–	?
c.	horn core	1		XIB.3-6	II	6	–	–	–	+
d.	horn core	1		LXXVII.9	II	6	+	–	–	–
e.	horn core		1	XIG/H.4-6.LXX	II	6	–	–	–	+
	horn core		1	XIC.4	II	5-6	–	–	–	?
f.	horn core	1		s.c.(3)	–	–	–	–	–	–
	horn core	1		s.c.	–	–	–	–	–	+
g.	horn core		1	s.c.	–	–	–	–	–	+
h.	horn core		1	s.c.	–	–	–	–	–	+
i.	horn core		1	s.c.	–	–	–	–	–	+
j.	horn core		1	s.c.	–	–	–	–	–	–
k.	atlas	1		CCXIII.a.10	II	6	+	–	–	–
l.	atlas	1		XIG/H.8/9.XXV	II	6	+	–	–	+
m.	scapula	1		D.3/1	III	3-4	–	–	–	?
n.	scapula	1		s.c.	–	–	–	–	–	+
o.	scapula		1	s.c.	–	–	–	?	–	–
p.	humerus	1		s.c.	–	–	–	–	?	–
q.	radius	1		s.c.	–	–	–	?	–	–
r.	radius		1	XIK.2-4	II-III	5	–	–	–	+
s.	ulna		1	s.c.	–	–	–	–	+	–
	metacarpal	1		s.c.	–	–	–	–	–	+
t.	metacarpal		1	XIH.2(25)	II-III	5	–	–	+	?
u.	metacarpal		1	CC.9	II	6	–	–	–	?
v.	metacarpal		1	XIG/H.20.XX	I	8	–	–	?	+
w.	pelvis		1	XIC.4	II	5-6	–	–	–	–
x.	pelvis		1	s.c.	–	–	–	+	–	+
y.	femur	1		CLXXI(6)/3	II	6	?	+	–	?
z.	tibia	1		XX.25	I	9	–	–	–	+
aa.	tibia	1		s.c.	–	–	–	?	?	+
bb.	calcaneum	1		CXXIII(2/3)	IV	1	+	–	–	–
cc.	calcaneum	1		XIM(2)CCI	II	5-6	–	–	–	?
dd.	calcaneum		1	XIB.1	II-III	5	–	–	–	–
ee.	calcaneum		1	s.c.	–	–	–	–	–	?
ff.	metatarsal	1		XIE.12.LVIII	II	7	+	–	–	–
gg.	metatarsal		1	CXXIV(4)	II	5-6	–	–	–	–
hh.	metatarsal		1	s.c.	–	–	–	?	?	–
ii.	first phalanx	1		CXX.4	IV	1	–	–	–	–

1) Notes: a = marks of butchering
           b = marks of intentional working
           c = marks of gnawing
           d = marks of burning
           + = present
           – = absent
           ? = uncertain
2) identification letters used in Tables 3 and 4.
3) s.c. = undefined provenience

FIGURE 3. *Possible representations of* Gazella *on painted Buff Ware sherds, datable to Shahr-i Sokhta Phases 6 or 5.*
  A. *provenience RYM.1-6*   B. 1. *provenience CC.6*
                                 2. *provenience RYL.6*
                                 3. *provenience XIX.5*
                                 4. *surface find*

TABLE 3

Skull measurements of Shahr-i Sokhta gazelles (in millimeters)

Neurocranium	a.*
width at the mastoid process	69.5
condylar width	42.6
inion - basion distance	36.5
basioccipital - parietal distance	55.7
distance between the horn cores	22.7
distance between the supraorbital foramina	27.5
1. maximum length of the dorsal edge of the horn core	151.0
2. minimum diameter at the base of the horn core	21.3
3. maximum diameter at the base of the horn core	30.4

Horn Cores	b	c	d	e	f	g	h	i	j
1.	190.0	–	165.0	(170.0)	–	–	–	–	–
2.	24.5	23.1	22.8	(20.0)	24.7	25.1	21.6	(24.5)	(24.8)
3.	33.0	31.0	30.3	–	33.5	34.2	29.2	–	(32.6)

* see Table 2 for identification of provenience.

TABLE 4

Post-cranial measurements for Shahr-i Sokhta and other gazelles (in millimeters)

Atlas	k*	l	uu	tt		
maximum length	–	47.0	60.5	55.0		
maximum height	32.7	27.7	32.0	33.2		
width of anterior artic. surface	41.8	37.8	44.6	41.5		
height of anterior artic. surf.	23.3	20.4	23.5	–		
width of posterior artic. surf.	37.0	34.2	38.6	37.0		
height of posterior artic. surf.	23.3	–	24.3	–		

Scapula	m	n	o	uu	vv	(1)	tt
length ventral angle	32.7	–	32.3	32.4	29.0	31.0	31.6
width glenoid cavity	22.8	22.1	24.8	24.9	21.2		23.5
minimum length neck	16.8	16.2	17.1	19.3	16.5	17.0	16.2
spine – glenoid cavity	15.7	16.0	16.3	21.0	16.5	18.5	15.5

Humerus	p	uu	vv	(2)	tt
distal diameter (ant.-post.)	(23.7)	25.5	22.8	–	23.8
distal width	25.0	27.6	26.0	31.0	26.8

Radius	q	r	uu	vv	(2)	tt
maximum length	160.0	–	170.0	160.0	160.0	155.0
proximal width	(27.2)	–	28.3	26.5	28.0	26.8
proximal diameter (ant.-post.)	15.6	–	15.5	16.0	16.0	14.5
distal width	24.8	22.9	25.3	23.5	26.0	24.2
distal diameter (ant.-post.)	19.0	16.6	19.1	17.5	–	18.7
minimum shaft width	(15.4)	–	16.4	14.8	16.0	14.0
minimum shaft diameter (a.-p.)	10.0	–	10.3	–	10.0	–

Ulna	s	uu	vv	(2)	tt
olecranon top – radius artic.	(37.2)	46.7	42.0	47.0	–
maximum diameter olecranon (a.-p.)	(19.1)	22.3	22.5	23.0	–
minimum diameter olecranon (a.-p.)	18.3	19.5	18.8	19.0	16.2
minimum width (med.-lat.)	6.2	7.1	6.7	7.0	5.2

Metacarpal	t	u	v	uu	vv	tt
minimum shaft diameter (a.-p.)	9.1	8.9	9.3	9.5	7.9	8.9
width distal articular surface	20.8	20.3	19.8	20.7	20.0	20.8
max. diam. distal artic. surf. (ant.-post.)	16.3	16.8	15.8	16.9	15.2	16.5

TABLE 4
(continued)
Post-cranial measurements for gazelles

Pelvis	w	x	uu	vv	tt		
minimum width ilium	9.0	–	–	8.7	8.0		
minimum depth ilium	18.0	–	–	17.8	16.3		
minimum width ischium	5.8	6.6	6.3	6.0	5.2		
minimum depth ischium	(19.4)	19.0	20.2	19.1	18.6		
length acetabulum	26.7	–	27.5	25.8	25.5		

Femur	y	uu	vv	(1)	(2)	(3)	tt
maximum prox. width	48.5	50.0	43.0	49.0	50.0	–	46.5
width articular head	26.4	26.5	23.0	26.0	24.0	–	25.0
diameter head (a.-p.)	19.9	19.4	18.2	–	–	20.0	18.8

Tibia	z	aa	uu	vv	(2)	tt	
proximal diameter (a.-p.)	40.0	38.7	43.8	39.0	–	40.0	
proximal width	38.3	(38.2)	38.9	36.0	40.0	38.8	

Calcaneum	ee	dd	bb	cc	uu	vv	(2)	tt
maximum length	59.0	59.0	61.4	58.2	62.6	55.0	64.0	57.2
min. width corpus	7.8	7.4	8.6	7.8	7.9	8.3	8.0	7.2
max. width dist. end	19.5	20.0	21.5	22.3	19.7	19.0	21.0	20.5
max. dorsal-plantar depth	23.7	–	24.2	23.7	24.3	21.7	24.0	22.3

Metatarsal	gg	hh	ff	uu	vv	(1)	tt
min. shaft diam. (a.-p.)	9.9	10.5	–	10.3	10.0	–	9.6
width dist. artic. surf.	21.3	21.8	20.7	21.6	20.7	20.5	21.9
max. diam. dist. artic. surf. (a.-p.)	16.9	16.9	16.8	17.3	–	16.0	17.0

Phalanx 1	ii	uu	vv	(1)	tt
maximum length	42.1	47.2	40.6	40.0	38.5

* see Table 2 for identification of provenience of Shahr-i Sokhta specimens
  see Table 1 for identification of comparative specimen
  the following measurements were calculated from illustrations:
  (1) = G. subgutturosa measurements from plate 26 of Boessneck and Krauss (1973)
  (2) = G. subgutturosa measurements from plates 25, 45, 69, 99 of Gromova (1950) and figure 53 of Gromova (1960)
  (3) = G. subgutturosa measurements from figure 123 of Hole et al. (1969)

the analysis of the remains from Shahr-i Sokhta, however, there do not seem to be any bones belonging to such another form included among the material. On the one hand, the fact that a second form has not been mentioned from amongst the faunal remains of the numerous archaeological sites in Iran and surrounding areas makes it seem possible that the introduction of this form took place in relatively recent times. On the other hand, Brian O'Reagan (personal communication), on the basis of his fieldwork among the gazelles of Iran, believes that the goitered gazelle adapts more readily to the presence of human settlement while the other form, which he classifies as Gazella dorcas fuscifrons Blanford, 1872, is a more decidedly desertic animal. Therefore, presuming that the dorcas gazelle had already reached Iran by the third millennium B.C., the species more likely to have been encountered by the hunters of Shahr-i Sokhta would have been the goitered gazelle.

References

Boessneck, J. and R. Krauss
   1973   "Die Tierwelt um Bastam/Nordwest-Azerbaidjan," *Archaeologische Mitteilungen aus Iran*, n.F., vol. 6, pp. 113-133. Berlin.

Duerst, J.U.
   1908   "Animal remains from the excavation at Anau and the horse of Anau in its relation to the history and to the races of domesticated horses," in R. Pumpelly, editor, *Explorations in Turkestan, Expedition of 1904*, pp. 341-442. Washington.

Firouz, E.
   1974   *Environment Iran*. Tehran.

Gromova, V.I.
   1950   *Opretelitel' mlekopitajuščih SSSR po kostjam skeleta*, vol. 1. Trudy Komissii po izučeniju cetvierticnogo perioda, vol. 9. Moskva-Leningrad.
   1960   *Opretelitel' mlekopitajuščih SSSR po kostjam skeleta*, vol. 2. Trudy Komissii po izučeniju cetvierticnogo perioda, vol. 16. Moskva-Leningrad.

Hilzheimer, M.
   1941   *Animal Remains from Tell Asmar*. Studies in Ancient Oriental Civilization, no. 20. Chicago.

Hole, F., K.V. Flannery, and J.A. Neely
   1969   *Prehistory and human ecology of the Deh Luran Plain. An Early Village Sequence from Khuzistan, Iran*. Memoirs of the Museum of Anthropology, University of Michigan, vol. 1. Ann Arbor.

Lange, J.
   1972   "Studien an Gazellenschadeln. Ein Beitrag zur Systematik der kleineren Gazellen, *Gazella* (de Blainville, 1816) *Säugetierkundliche Mitteilungen*, vol. 20, pp. 193-249.

Lay, D.M.
   1967   *A Study of the Mammals of Iran, resulting from the Street Expedition of 1962-1963*. Fieldiana: Zoology, vol. 54. Chicago.

Lydekker, R.
   1910   "The Gazelles of Seistan," *Nature*, pp. 201-202.

Misonne, X.
   1959   *Analyse zoogéographique des mammifères de l'Iran*. Mémoires de l'Institut Royal des Sciences Naturelles de Belgique, deuxième série, fasc. 59. Bruxelles.

THE BONE REMAINS OF SMALL WILD CARNIVORES FROM SHAHR-I SOKHTA

Lucia Caloi, Istituto di Geologia e Paleontologia
University of Rome

Lutra lutra Linné, 1758 (Tables 1 and 2)

The otter is represented by only two bones: a complete right ulna [provenience "s.c."], which lacks only a small splinter from its posterior border (Fig. 1B), and a left metatarsal V [provenience LIV.12]. The latter comes from Period II. These remains have been compared with a skeleton from Italy (*Lutra lutra*, male, no. 882, Collez. Cardini, Istituto Italiano di Paleontologia Umana, Rome). The Shahr-i Sokhta ulna is slightly smaller but very similar in shape and proportions to the Italian one while our metatarsal is straighter and longer with a wider proximal end and shaft. In addition the metatarsal from Shahr-i Sokhta bears a short ridge, which is missing in the Italian specimen, on the plantar face of the shaft immediately above the distal articular surface almost continuous with the intertrochlear crest.

At present, in Sistan, is found a form of common otter identified as *Lutra lutra seistanica* Birula. Since the lacustrine basin of Sistan has persisted from the time of the prehistoric settlement at Shahr-i Sokhta to the present, it is likely that the archaeological remains belong to the same form.

Vulpes vulpes Linné, 1758 (Tables 1 and 3)

The red fox is represented only in Shahr-i Sokhta Period II and only by a single specimen, a left mandible [provenience XXVIII.4] which bears traces of burning (Fig. 1A). The coronoid process and the portion in front of the P_1 are missing; of the teeth only the M_2 remains. The mandible is similar in size to that of an average sized red fox (condylo-incisive length of 97-126 mm, Toschi 1965) but is smaller than the largest of the large Italian subspecies *Vulpes vulpes crucigera* (condylo-incisive length of 97-119 mm, Toschi 1965). The horizontal ramus of our specimen is not very high and the height decreases in an oral direction more rapidly than in the Italian examples. In addition, the coronoid process of our specimen is less broad at the base than in the Italian material. Its condyle is nearly horizontal instead of being inclined inwards and the tooth alveoli are regularly ranged, there being only a small overlapping between the alveoli of P_4 and M_1. The M_2 appears longer and narrower than in the Italian specimens as well as narrower between the external posterior tubercle and the two anterior ones. The mandible from Tepe Ali Kosh in southwest Iran (6750-6000 B.C.) shown in Figure 132 of Hole, Flannery, and Neely (1969) appears much smaller than our specimen and also has a relatively higher horizontal ramus and a longer M_2.

The extreme rarity of fox remains from Shahr-i Sokhta is probably linked to urbanization at the site since a town does not provide a suitable habitat for this wild species, which in turn was not and could not be an important source of food.

Felis cf. *catus* Linné, 1758 (Tables 1 and 4)

The cats are represented at Shahr-i Sokhta by a mandible (Fig. 1C) and a humerus (Fig. 1D). Both bones clearly belong to animals of the genus *Felis* on the basis of size and shape and probably come from the wild or steppe cat of the "ornata group" (Grzimek 1973, vol. 12).

The mandible [provenience CCLXV.2] is from Shahr-i Sokhta Period II. The bone is fairly robust and of a size comparable to the medium-size wild cat "*Felis silvestris*." The horizontal ramus is not very high and the symphysis is short as in the comparative specimens of wild cat and unlike the longer domestic condition. While only a fragment of the carnassial remains, the length of the tooth row, to judge from the alveoli, is remarkable. In particular, the alveolus of the canine is very large and the fragment of the carnassial suggests a well-developed tooth. The distance between the canine and the first premolar is very short. The coronoid process is low and short and the coronoid fossa appears very deep because of the lateral folding of the anterior border of the ascending ramus immediately behind the carnassial. The condyle is short and of average thickness.

Features like the large teeth and short diastema support the identification of our specimen as a wild cat. The extreme rarity of such finds at Shahr-i Sokhta provides indirect evidence that the cat was not commonly living in association with humans at the site. In addition no clear

TABLE 1

Bone remains of wild carnivores from Shahr-i Sokhta

Species:	Element:	# of specimens rt.	lft.	Provenience	Period	Phase
Lutra lutra						
	ulna	1		s.c.[1]	-	-
	metatarsal V		1	LIV.12	II	7
Vulpes vulpes						
	mandible		1	XXVIII.4	II	6
Felis cf. *catus*						
	mandible	1		CCLXV.2	II	6
	humerus	1		RYL.3	II-III	5

1) s.c. = undefined provenience

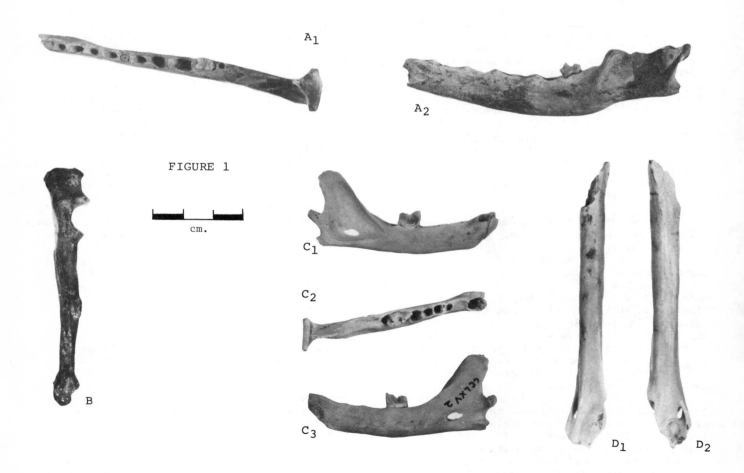

FIGURE 1

TABLE 2: *Lutra lutra* measurements
(in millimeters)

metatarsal V:

maximum length	39.0
breadth of prox. epiphysis	8.0
minimum breadth of shaft	3.5
breadth of dist. artic. surf.	6.2

ulna:

maximum length	80.7
height of trochlear notch	10.3

TABLE 4: *Felis* cf. *catus* measurements
(in millimeters)

mandible:	length (condyle – I_1)	63.0
	height of ramus at M_1	10.0
	thickness of ramus at M_1	6.0
	height of ascending ramus	(26.0)
	alveolar length P_3–M_1	24.0
	alveolar length P_3	6.5
	P_4	8.0
	M_1	9.5
humerus:	breadth of shaft	8.7
	height of epitrochlear foramen and anterior groove	10.3

TABLE 3: *Vulpes vulpes* measurements
(in millimeters)

mandible:	height of ramus behind P_1	10.8
	height of ramus behind M_1	13.8
	alveolar length P_1–M_3	61.7
	alveolar length P_1	4.5
	P_2	8.5
	P_3	9.2
	P_4	9.0
	M_1	15.0
	M_2	7.2
	M_3	3.0
	crown length M_2	7.5
	crown breadth M_2	5.9

FIGURE 1. *Bones of small wild carnivores from Shahr-i Sokhta (scale in centimeters is approximate).*
 A. *left mandible of Vulpes vulpes [provenience XXVIII.4]*
  1. *occlusal view*
  2. *lateral view*
 B. *right ulna of Lutra lutra [s.c.], medial view*
 C. *right mandible of Felis cf. catus [CCLXV.2]*
  1. *lateral view*
  2. *occlusal view*
  3. *medial view*
 D. *right humerus of Felis cf. catus [RYL.3]*
  1. *caudal view of distal shaft*
  2. *cranial view of distal shaft*

osteological evidence for the existence of the domestic cat at a comparable period has been found. The mandible identified as *Felis* cf. *libyca* from Ali Kosh (6750-6000 B.C.) and illustrated in Figure 132 of Hole, Flannery, and Neely (1969) is considerably smaller than our specimen.

The right humerus from Period II-III [provenience RYL.3] includes only the shaft. Although at present only the wild cat lives in Sistan, it is possible that in past times other cats, now confined to neighboring regions, were living there. The dimensions of our specimen, however, are too large for *Felis manul* and too small for *Felis chaus* while *Felis thinobius* shows a different morphology of the distal extremity (Gromova 1950, Figs. 58,59). In the humerus of the last named species, the bony bar that delimits internally the epithroclear foramen appears thicker and the foramen less developed than in *Felis catus* and the shaft is more bent. In our specimen the diaphysis is almost straight and the fossa for the olecranon appears very large with its upper border gently inclined downwards instead of forming a sudden step as in wild and domestic specimens. The bony bar divaricates very little while the suprathroclear foramen is very pronounced with the grooves that prolong it on both the dorsal and ventral faces appearing long and deep; the groove in the dorsal face extends beyond the muscle scar more than in the comparative material.

References

Gromova, V.I.
 1950 *Opretelitel' mlekopitajuščih SSSR po kostjam skeleta*, vol. 1. Trudy Komissii po izučeniju cetvierticnogo perioda, vol. 9. Moskva-Leningrad.
Grzimek, B. (editor)
 1973 *Vita degli animali*, vol. 12 (Mammiferi 3). Varese.

Hole, F., K.V. Flannery, and J.A. Neely
  1969   *Prehistory and human ecology of the Deh Luran Plain. An Early Village Sequence from Khuzistan, Iran.* Memoirs of the Museum of Anthropology, University of Michigan, vol. 1. Ann Arbor.

Toschi, A.
  1965   "Mammalia," in *Fauna d'Italia*, vol. 7. Bologna.

PART FIVE

CODING SYSTEMS

One feature of the 1970's has been the explosive growth of computer processing applications to all fields of academic endeavor and zooarchaeology is no exception. Faunal analysis requires that large quantities of data characterizable by multiple attributes each with many states be manipulated for purposes of quantification and pattern recognition. In some ways, the basic unit of faunal studies - the bone (when dealing with vertebrate fauna) - is particularly well suited for computer coding because the possible states of many of the attributes are definable in discrete fachion based on zoological criteria. Not surprisingly, it is in these aspects of identification of a bone as to species, body part (element), side (right/left), state of fusion, and so on that the various coding systems are the most similar. Systems differ the most from one another when dealing with descriptions of the results of cultural processes which have acted on zoological material, e.g., burning, fragmentation, and butchering. For such attributes variation will be of continuous rather than of a discrete nature and attempts to codify them necessarily will be more subjective and tailored to the individual interests of the analyst.

The three coding systems included in this part all developed from the same ancestor, the system originally put together at Çayönü in 1970 (Reed 1971). The varying final states of these codes are partly a result of varying degrees of interaction between their creators, but mostly the result of differing perceptions of needs, different working styles, and different interests. Uerpmann's code reflects his interest in using a coding system to record basic zoological information—both attribute and metrical. The specimen serves as the unit of analysis and the attempt is made to enter all data for a piece on one data line. Meadow's code also is based on the specimen, but because he wanted his code to accommodate a great deal of cultural information along with the zoological data, he found it necessary to separate into two data lines discrete attributes and measurements. In so doing, however, unique labeling of each specimen became necessary for purposes of cross-referencing. Redding, Zeder, and McArdle's code also requires the use of more than one data line per specimen, but instead of the piece (or 'bone'), they chose the element as the basic unit for coding so that it would be possible to provide detailed descriptions of fragmentation and ageing criteria primarily for skull elements. For example, the cheek area of the mandible of a bovid with a full set of adult teeth (PM_{2-4} and M_{1-3}) would be coded by Uerpmann and Meadow using one data line (with an additional line or lines for measurements). Redding, Zeder, and McArcle would use seven lines to record the same information, one per tooth (each an 'element' just as each skull bone would be) and one for the jaw itself. Here the unique labeling of each element is necessary to unify all those belonging to one specimen. The labels are also used to cross-reference the descriptive data with measurements recorded separately by another system (not published here).

The codes of Redding, Zeder, and McArdle and of Meadow show a particular emphasis on the recording of cultural information. Both codes provide categories that permit detailed descriptions of burning, fragmentation, and cultural modification. Meadow's code allows for the recovery of information on types of fracture and location of butchering marks.

These comments have pointed out only some of the more salient differences between the three numerical coding systems. By presenting them together here, we hope to provide the would-be user with an opportunity to evaluate the strengths and limitations of each and to either adopt one outright or, ideally, to pick and choose and modify--to develop his or her own system compatible with individual requirements. There are, of course, other systems such as those described by Clutton-Brock (1975), Gifford and Crader (1977), and LaBianca (noted in Part One). The basic requirement of any approach, whether computerized or not, is that certain types of zooarchaeological information be recorded in a format which permits its retrieval and use by the investigator and by others to whom the data is made available.

References

Clutton-Brock, J.
    1975   "A system for the retrieval of data relating to animal remains from archaeological sites," in A.T. Clason, editor, *Archaeozoological Studies,* pp. 21-34. Amsterdam/New York.

Gifford, D.P. and D.C. Crader
    1977   "A computer coding system for archaeological faunal remains," *American Antiquity,* vol. 42, pp. 225-238.

Reed, C.A.
    1971   "New method for recording and analyzing faunal material from archaeological sites," manuscript circulated at the Section on Animal Domestication of the Third International Congress of the Museums of Agriculture at Budapest, 19-23 April, 1971.

R.H.M.
M.A.Z.

"BONESORT II" - A SYSTEM FOR THE COMPUTER PROCESSING OF IDENTIFIABLE FAUNAL MATERIAL

Richard W. Redding, Museum of Zoology
University of Michigan
Melinda A. Zeder, Museum of Anthropology
University of Michigan
John McArdle, Illinois Wesleyan University
Bloomington

Introduction

Zooarchaeology has in the past lacked comprehensive and consistent information recording systems. The wide diversity of investigative methods has tended to hamper assessment of similarities and differences among faunal remains from archaeological assemblages. The increase in both quantity and quality of archaeological faunal collections together with the growth in the complexity of the problems being investigated makes apparent the need for systems which facilitate the recording of certain basic kinds of information using numerical coding and computer processing.

Our present approach to recording and analyzing data has developed from the work of several individuals. As a result of fieldwork in 1970 with Charles Reed in southeastern Turkey and conversations with archaeologists at the excavations, McArdle devised a preliminary scheme for numerically coding morphological attribute data in a comprehensive manner. The aim of this system was to maximize the amount of information derived from the raw data while minimizing the amount of time expended on each assemblage. During the field season, these preliminary attempts were discussed with Redding who also was interested in developing such a system. Subsequent to their meeting, Redding and McArdle returned to their respective universities (Michigan and Chicago) and independently continued their work. The parallel evolution of ideas produced two essentially similar methods as demonstrated in papers by McArdle (1977) and Redding, Pires-Ferreira, and Zeder (1977).

In 1975, Redding, Zeder, and McArdle collaborated on the development of the system presented here. This system includes a comprehensive numerical code which allows the recording of information on bone location, type, fragmentation, and cultural modification. A sorting program is also included which converts numerical data into abbreviated verbal form and facilitates examination of the data to permit recognition of patterning. BONESORT II also provides counts of various combinations of attributes. Further manipulation of the data can be effected through the application of a number of packaged statistical programs. Redding and Zeder also employ separate systems for the recording of measurement data of identifiable material and for the analysis of "unidentifiable" specimens.

Since the actual value of any information derived from a faunal study is, to a great degree, a function of the approach employed by the individual researcher, there is a critical need for agreement on the recording of specific descriptive attributes, both biological and cultural. A system such as ours maximizes recovery of information from faunal material and enhances comparability of the analyses.

References

McArdle, J.
   1977   "A numerical (computerized) method for quantifying zooarchaeological comparisons," *Paléorient*, vol. 3.

Redding, R.W., J. Wheeler Pires-Ferreira, M. Zeder
   1977   "A proposed system for computer analysis of identifiable faunal material from archaeological sites," *Paléorient*, vol. 3.

```
C
C
C                          BONESORT II
C
C     THIS IS A COMPUTER PROGRAM WHICH SELECTS DATA FROM CODED FAUNAL
C ANALYSES ACCORDING TO ANY COMBINATION OF 37 POSSIBLE CRITERIA AND
C PRESENTS THE DATA IN ABBREVIATED VERBAL FORM. THE CODE USED BY THE
C AUTHORS IS PRESENTED AND EXPLAINED BELOW. THE SORT PROGRAM (BONESORT
C II) IS ALSO PRESENTED WITH STEP BY STEP COMMENTS EXPLAINING ITS OP-
C PERATION. FINALLY A COPY OF THE TRANSLATION FILE (VERB) USED IN BONE-
C SORT II, WITH COMMENTS, IS INCLUDED.
C
C    ********************THE CODE*******************************
C
C 01.) SITE: (COLUMNS 1-3) THIS VARIABLE IS AN ASSIGNED NUMERIC CODE
C FOR A GROUP OF FAUNAL MATERIAL. AT MICHIGAN THIS NUMBER IS ASSIGN-
C ED TO MATERIAL BY SITE. THIS NUMBER IS USED IN REFERENCING THIS
C MATERIAL IN OUR CATALOGUE SYSTEM AND IN REFERENCING ANY PUBLICATION
C PERTAINING TO THIS MATERIAL.
C
C 02.) ELEMENT NO: (COLUMNS 4-8)  NUMBER OF THE BONE FROM 00001 TO
C      99999.
C 03.) STRATIGRAPHIC CODE I: (COLUMNS 9-12)
C 04.) STRATIGRAPHIC CODE II: (COLUMNS 13-16)
C 05.) STRATIGRAPHIC CODE III: (COLUMNS 17-20)
C THE POSITION OF THE ELEMENT WITHIN THE SITE CAN BE RECORDED BY
C MODIFYING THE GRID SYSTEM OF THE EXCAVATOR INTO THESE 3 VARIABLES.
C
C 06.) PERIOD: (COLUMNS 21-22) THIS VARIABLE REFERS TO THE PERIOD
C DESIGNATION ASSIGNED BY THE INVESTIGATOR TO THE PROVENIENCE UNIT
C IN WHICH THE BONE WAS FOUND. THIS LIST MAY BE MODIFIED TO FIT
C THE CULTURAL SEQUENCES IN DIFFERENT AREAS.
C
C INDETERMINATE: 01, INDT       MIDDLE ELAMITE:     09, MELM
C MUSHKI:        02, MUSH       LATE ELAMITE:       10, LELM
C JARI:          03, JARI       PARTHO-SASSANIAN:   11, PRSA
C BAKUN:         04, BKUN       ISLAMIC:            12, ISLM
C LAPUI:         05, LAPU
C BANESH:        06, BNSH
C KAFTARI:       07, KAFT
C QALEH:         08, QLEH
C
C 07.) CONTEXT: (COLUMNS 23-24) THE ARCHEOLOGICAL CONTEXT IN WHICH
C THE BONE WAS FOUND IS CODED HERE.
C
C INDETERMINATE:01, INDT        DUMP:               05, DUMP
C FLOOR:        02, FLOR        FILL:               06, FILL
C FEATURE:      03, FEAT
C PIT:          04, PIT
C
C 08.) CLASS: (COLUMN 25) THE TAXONOMIC CLASS OF ANIMAL REPRESENTED BY
C THE BONE IS CODED HERE.
C
C INDETERMINATE: 1, INDT        AMPHIBIA:   5, AMPH
C INVERTEBRATE:  2, INVT        REPTILIA:   6, REPT
C CHONDRICHTHYES:3, CHON        AVES:       7, AVES
C OSTEICHTHYES:  4, OSTE        MAMMALIA:   8, MAMM
C
C 09.) IDENTIFICATION: (COLUMNS 26-28) THE VERTEBRATE REPRESENTED BY
C THE ELEMENT IS RECORDED IN THIS VARIABLE. A LIST OF ANIMALS AND
C THE CODES FOR THEM MUST BE PREPARED FOR EACH GEOGRAPHIC AREA. THE
C LIST BELOW HAS BEEN PREPARED FOR IRAN AND IRAQ.
C
C INDETERMINATE:         001, INDT    LARGE CARNIVORE:      032, LCRN
C LARGE MAMMAL:          002, LMAM    MEDIUM CARNIVORE      033, MCRN
C MEDIUM MAMMAL:         003, MMAM    SMALL CARNIVORE:      034, SCRN
C SMALL MAMMAL:          004, SMAM    VIVERRID:             035, VIVR
C LARGE ARTIODACTYL:     005, LART    HERPESTES SP.:        036, HRPS
C MEDIUM ARTIODACTYL:    006, MART    H. AUROPUNCTATUS:     037, HAUR
C SMALL ARTIODACTYL:     007, SART    H. ICHNEUMON:         038, HICH
C BOVID:                 008, BOVD    H. EDWARDSI:          039, HEDW
C OVIS-CAPRA-GAZELLA:    009, OCG     ICHUNEMIA ALBICAUDA:  040, ICAL
C OVIS-CAPRA:            010, OVCP    GENETTA GENETTA:      041, GEGE
C OVIS:                  011, OVIS    HYAENA:               042, HYNA
C CAPRA:                 012, CPRA    LARGE FELID:          043, LFEL
C GAZELLA SP.:           013, GZLA    MEDIUM FELID:         044, MFEL
C G. DORCAS:             014, GDOR    SMALL FELID:          045, SFEL
C G. GAZELLA:            015, GGAZ    ACINONYX              046, ACIX
C G. SUBGUTTUROSA:       016, GSUB    CARACAL:              047, CARA
C BOS:                   017, BOS     FELIS CATUS:          048, FCAT
C BUBULUS:               018, BUBU    F. CHAUS:             049, FCHA
C CERVID:                019, CERV    F. LEO:               050, FLEO
C CAPREOLUS:             020, CPRL    F. MARGARITA:         051, FMAR
C CERVUS:                021, CRVS    F. PARDUS:            052, FPAR
C DAMA:                  022, DAMA    F. UNCIA:             053, FUNC
C SUID:                  023, SUID    LYNX:                 054, LYNX
C SUS SCROFA:            024, SSCR    MEDIUM MUSTELID:      055, MMUS
C CAMELID:               025, CMLD    SMALL MUSTELID:       056, SMUS
C CAMELUS:               026, CMLS    LUTRA:                057, LTRA
C EQUID:                 027, EQUD    MARTES FOINA:         058, MFOI
C E. ASINUS/HEMIONUS:    028, EQAH    MELES MELES:          059, MEME
C E. ASINUS:             029, EQAS    MELLIOVORA CAPENSIS:  060, MCAP
C E. EQUUS:              030, EQEQ    MUSTELA NIVALIS:      061, MNIV
C E. HEMIONUS:           031, EQHM    VORMELA PEREGUSNA:    062, VPER
```

```
C LARGE CANID:        063, LCAN       MESOCRICETUS        092, MESO
C MEDIUM CANID:       064, MCAN       MICROTINE:          093, MICR
C SMALL CANID:        065, SCAN       ARVICOLA:           094, ARVI
C CANIS SP.:          066, CNIS       ELLOBIUS:           095, ELLO
C C. AUREUS:          067, CAUR       MICROTUS:           096, MCRT
C C. FAMILIARIS:      068, CFAM       GERBELLINE:         097, GRBL
C C. LUPUS:           069, CLUP       GERBILLUS:          098, GRBS
C FENNECUS:           070, FENN       MERIONES:           099, MRIO
C VULPES SP.:         071, VLPS       PSAMMOMYS:          100, PSAM
C V. CANA:            072, VCAN       SEKEETAMYS:         101, SEKE
C V. RUPPELLI:        073, VRUP       TATERA:             102, TATR
C V. VULPES:          074, VVUL       MURID:              103, MURD
C URUS:               075, URUS       ACOMYS:             104, ACOM
C PROCAVIA:           076, PROC       APODEMUS:           105, APOD
C ERINACEID:          077, ERND       ARVICANTHIS:        106, ARVI
C ERINACEUS:          078, ERNC       MUS:                107, MUS
C HEMIECHINUS:        079, HMIC       NESOKIA:            108, NESO
C PARAECHINUS:        080, PRCH       RATTUS:             109, RATT
C TALPA:              081, TALP       DIPODID:            110, DIPO
C SORICID:            082, SORI       ALLACTAGA:          111, ALLA
C CROCIDURA:          083, CROC       JACULUS:            112, JACU
C NEOMYS:             084, NEOM       GLIRID:             113, GLIR
C SUNCUS:             085, SUNC       DRYOMYS:            114, DRYO
C CHRIOPTERID:        086, CHIR       GLIS:               115, GLIS
C RODENT:             087, RODT       SPALAX:             116, SPAL
C SCIURUS:            088, SCUI       HYSTRIX:            117, HYST
C CRICETID:           089, CRIC       LAGOMORPH:          118, LAGO
C CALOMYSCUS:         090, CALO       LEPUS:              119, LEPU
C CRICETULUS:         091, CRCT       OCHOTONA:           120, OCHO
C
C 10.) RELIABILITY OF IDENTIFICATION: (COLUMN 29) A SUBJECTIVE
C ESTIMATE OF THE RELIABILITY OF THE IDENTIFICATION IS CODED IN
C THIS VARIABLE.
C
C INDETERMINATE:     1, INDT         POSSIBLY RELIABLE:  3, PREL
C RELIABLE:          2, RELI         NOT RELIABLE:       4, NREL
C
C 11.) ECOLOGICAL INDEX: (COLUMNS 30-31) A GROSS CLASSIFICATION
C OF THE ECOSYSTEM OF WHICH THE ANIMAL IDENTIFIED IS A MEMBER IS
C INDICATED HERE. THIS VARIABLE IS NOT USED AT MICHIGAN.
C
C 12.) ELEMENT: (COLUMNS 32-34) THE TYPE OF ELEMENT IS CODED HERE. ALL
C COMPLETE OR PARTIAL SKULLS HAVE ELEMENTS RECORDED INDIVIDUALLY.
C ALL ARTICULATED ELEMENTS, INCLUDING TEETH IN MANDIBLES AND MAXILLAE.
C ARE INDIVIDUALLY CODED.
C
C INDETERMINATE:      001, INDT      SCAPULA:            048, SCAP
C BASIOCCIPITAL:      002, BASO      CLAVICLE:           049, CLAV
C OCCIPITAL:          003, OCIP      HUMERUS:            050, HUMR
C BASISPHENOID:       004, BASP      RADIUS:             051, RADI
C MESETHMOID:         005, MESE      ULNA:               052, ULNA
C PRESPHENOID:        006, PRSP      RADIAL CARPAL:      053, RCRP
C ALISPHENOID:        007, ALIS      INTERMED. CARPAL:   054, ICRP
C PTERYGOID:          008, PTER      ULNAR CARPAL:       055, UCRP
C VOMER:              009, VOMR      CENTRAL CARPAL:     056, CCRP
C PALATINE:           010, PALA      PISIFORM:           057, PISI
C TURBINAL:           011, TURB      1ST CARPAL:         058, 1CRP
C SUPRAOCCIPITAL:     012, SUPR      2ND CARPAL:         059, 2CRP
C INTERPARIETAL:      013, INTP      3RD CARPAL:         060, 3CRP
C PARIETAL:           014, PARI      4TH CARPAL:         061, 4CRP
C FRONTAL:            015, FRON      RAD-INTERMED CARPAL:062. RICR
C PETROUS-TEMPORAL:   016, PSTP      2ND & 3RD CARPAL:   063, 23CR
C SQUAMOUS/TEMPORAL: 017, SQTM       METACARPAL I:       064, MTC1
C MALAR:              018, MLAR      METACARPAL II:      065, MTC2
C LACRIMAL:           019, LACR      METACARPAL III:     066, MTC3
C NASAL:              020, NSAL      METACARPAL IV:      067, MTC4
C PREMAX. WITH TEETH:021, PMWT       METACARPAL V:       068, MTC5
C PREMAX. WO. TEETH: 022, PMOT       METACARPAL III-IV:  069, MC34
C MAXILLIA W. TEETH: 023, MXWT       METACARPAL INDT.:   070, MTCR
C MAXILLIA WO. TEETH:024, MXOT       PROXIMAL SESAMOID:  071, PSEF
C MANDIBLE W. TEETH: 025, MNWT       DISTAL SESAMOID:    072, DSEF
C MANDIBLE WO. TEETH:026, MNOT       SEASMOID:           073, SESF
C ASSOCIATED TOOTH:   027, TOIA      1ST PHALANX:        074, 1PHF
C UNASSOCIATED TOOTH:028, TONA       2ND PHALANX:        075, 2PHF
C HORN CORE:          029, HORN      3RD PHALANX:        076, 3PHF
C ANTLER:             030, ANTL      PHALANX:            077, PHLF
C STYLOHYOID:         031, STYL      INNOMINATE:         078, INNO
C EPIHYOID:           032, EPIH      ILIUM-ISCHIUMM:     079, ILIS
C CERATOHYOID:        033, CERA      ILIUM-PUBIS:        080, ILPU
C BASIOHYOID:         034, BASH      ISCHIUM-PUBIS:      081, ISPU
C THYROHYOID:         035, THYH      ILIUM:              082, ILLI
C ATLAS:              036, ATLS      ISCHIUM:            083, ISHI
C AXIS:               037, AXIS      PUBIS:              084, PUBS
C CERVICAL VERTEBRA:  038, CVRT      OS PENIS:           085, OSPN
C THORACIC VERTEBRA:  039, TVRT      FEMUR:              086, FEMR
C LUMBAR VERTEBRA:    040, LVRT      PATELLA:            087, PATL
C SACRUM:             041, SACR      TIBIA:              088, TBIA
C CAUDAL VERTEBRA:    042, CAUD      FIBIULA:            089, FIBI
C RIB:                043, RIB       LATERAL MALLEOLUS:  090. LTML
C COSTAL RIB:         044, CRIB      ASTRAGALUS:         091, ASTR
C PRESTERNUM:         045, PRST      CALCANEUM:          092, CALC
C MESOSTERNUM:        046, MSST      CENTRAL TARSAL:     093, CTRS
C XIPHISTERNUM:       047, XIPH      1ST TARSAL:         094, 1TRS
```

```
C   2ND TARSAL:            095, 2TRS     SESAMOID:              108, SESH
C   3RD TARSAL:            096, 3TRS     1ST PHALANX:           109, 1PHH
C   4TH TARSAL:            097, 4TRS     2ND PHALANX:           110, 2PHH
C   CENTRAL&4TH TARSAL:    098, C4TR     3RD PHALANX:           111, 3PHH
C   METATARSAL I:          099, MTT1     PHALANX:               112, PHLH
C   METATARSAL II:         100, MTT2     METAPODIAL III-IV:     113, MP34
C   METATARSAL III:        101, MTT3     METAPODIAL INDT:       114, MTPD
C   METATARSAL IV:         102, MTT4     PROXIMAL SESAMOID:     115, PRSE
C   METATARSAL V:          103, MTT5     DISTAL SESAMOID:       116, DSSE
C   METATARSAL III-IV:     104, MT34     SESAMOID:              117, SEAS
C   METATARSAL INDT.:      105: MTTR     1ST PHALANX:           118, 1PHL
C   PROXIMAL SESAMOID:     106, PSEH     2ND PHALANX:           119, 2PHL
C   DISTAL SESAMOID:       107, DSEH     3RD PHALANX:           120, 3PHL
C                                        1 OR 2 PHALANX:        121, PHLX
C                                        2ND & 3RD TARSAL:      122, 2&3T
C
C   13.) SYMMETRY: (COLUMN 35)  THE SYMMETRY OF EACH ELEMENT IS RECORDED
C   HERE.
C
C   INDETERMINATE: 1, INDT           LEFT:   3, LEFT
C   RIGHT:         2, RGHT           MEDIAL: 4, MEDI
C
C   14.) FUSION: (COLUMNS 36-37) THE STATES OF FUSION OF EACH ELEMENT IS
C   CODED HERE. FUSION DATA, NOT AGE ESTIMATES, ARE USED SO THAT ONLY
C   RAW DATA, NOT SUBJECTIVE CONCLUSIONS, ARE RECORDED.
C
C   INDETERMINATE:            01, INDT   PROX FUSING DIST FUSING: 09, PIDI
C   FUSED:                    02, FUSD   PROX FUSING DIST UNFUSED:10, PIDU
C   FUSING:                   03, FUSI   PROX UNFUSED DIST FUSED: 11, PUDF
C   UNFUSED:                  04, UFUS   PROX UNFUSED DIST FUSING:12, PUDI
C   PROX FUSED DIST FUSED:    05, PFDF   PROX UNFUSED DIS UNFUSED:13, PUDU
C   PROX FUSED DIST FUSING:   06, PFDI   FOETAL:                  14, FOET
C   PROX FUSED DIST UNFUSED:  07, PFDU
C   PROX FUSING DIST FUSED:   08, PIDF
C
C   15.) FRAGMENT: (COLUMN 38)  A SUBJECTIVE CLASSIFICATION OF FRAG-
C   MENTATION IS CODED IN THIS VARIBLE.
C
C   INDETERMINATE:  1, INDT          1/2 TO 3/4 OF:   4, 3412
C   COMPLETE:       2, COMP          1/4 TO 1/2 OF:   5, 1412
C   3/4 TO COMPLETE: 3, 3/4C         LESS THAN 1/4 OF: 6, L1/4
C
C   16.) ORIGIN OF FRAGMENTATION. (COLUMN 39)  A CHARACTERIZATION OF
C   THE MAJOR SOURCE OF FRAGMENTATION EXHIBITED BY THE ELEMENT IS CODED
C   IN THIS VARIABLE.
C
C                  INDETERMINATE:                     1, INDT
C                  PRE-DEPOSITIONAL:                  2, PRED
C                  POSSIBLY THE ABOVE:                3, ?PRE
C                  POST-DEPOSITIONAL (ANCIENT):       4, POST
C                  POSSIBLY THE ABOVE:                5, ?PST
C                  RECENT (IN EXCAVATION OR SHIPMENT):6, RECT
C                  POSSIBLY RECENT:                   7, ?RCT
C
C   17.) FRAGMENTATION 1: (COLUMNS 40-41) IF AN ELEMENT IS INCOMPLETE, ITS
C   DEGREE OF FRAGMENTATION IS CODED HERE. THIS IS THE MAJOR VARIABLE
C   DESCRIBING FRAGMENTATION FOR ALL ELEMENTS EXCEPT TEETH AND VERTEBRA.
C   FOR LIMBS:
C   INDETERMINATE:         01, INDT     SHAFT:                06, SHFT
C   COMPLETE:              02, COMP     DISTAL END & SHAFT:07, DE&S
C   PROXIMAL END:          03, PEND     DISTAL SHAFT:         08, DSFT
C   PROXIMAL SHAFT         04, PSFT     DISTAL END:           09, DEND
C   PROXIMAL END & SHAFT:  05, PE&S
C
C   FOR MANDIBLES:(*=SAME TRANSLATION AS FOR LIMBS)
C   INDETERMINATE:         01, INDT     DIASTEMA & CHEEK:     07, *
C   COMPLETE:              02, COMP     SYMPHASIS TO CHEEK:08, *
C   ARTICULATION:          03, *        DIASTEMA TO ARTIC:    09, *
C   RAMUS:                 04, *        SYMPHASIS&DIASTEMA:10, SY&D
C   ARTICULATION & RAMUS:  05, *        SYMPHASIS:            11, SYMP
C   CHEEK & ARTICULATION:  06, *        DIASTEMA:             12, DIAS
C                                       CHEEK:                13, CHEK
C
C   FOR MAXILLA AND SKULLS:
C   INDETERMINATE:         01, INDT     CENTRAL:              04, *
C   COMPLETE:              02, COMP     DISTAL:               05, *
C   PROXIMAL:              03, PROX
C
C   18.) FRAGMENTATION 2: (COLUMN 42)  IF AN ELEMENT IS INCOMPLETE
C   ITS ANTERIOR/POSTERIOR PLANE IS DESCRIBED BY THIS VARIABLE. IT IS TO
C   BE USED WITH ALL ELEMENTS EXCEPT FOR THE SKULL AND MANDIBLE.
C
C   FOR LIMBS AND VERTEBRA:
C   INDETERMINATE: 1, INDT           POSTERIOR:   4, POST
C   COMPLETE:      2, COMP           CENTRAL:     5, CENT
C   ANTERIOR:      3, ANTR
C
C   FOR TEETH:
C   INDETERMINATE: 1, INDT           DISTAL:      4, *
C   COMPLETE:      2, COMP           CENTRAL:     5, CENT
C   MESIAL:        3, *
C
```

```
C 19.) FRAGMENTATION 3: (COLUMN 43) IF AN ELEMENT IS INCOMPLETE,
C ITS FRAGMENTATION IN TERMS OF LATERAL/MEDIAL RELATION TO THE BODY IS
C CODED HERE. THIS VARIABLE IS APPLICABLE TO ALL ELEMENTS. FOR
C VERTEBRA AND SOME SKULL ELEMENTS, LATERAL READS AS RIGHT HALF AND
C MEDIAL AS LEFT HALF.
C
C FOR LIMBS, MANDIBLES, MAXILLA, AND SKULLS:
C INDETERMINATE: 1, INDT           MEDIAL:          4, MEDI
C COMPLETE:      2, COMP           CENTRAL:         5, CENT
C LATERAL:       3, LAT
C
C FOR TEETH:
C INDETERMINATE: 1, INDT           LINGUAL:         4, *
C COMPLETE:      2, COMP           CENTRAL          5, CENT
C LABIAL:        3, *
C
C FOR VERRTEBRA:
C INDETERMINATE: 1, INDT           LEFT 1/2:        4, *
C COMPLETE:      2, COMP           CENTRAL:         5, CENT
C RIGHT 1/2:     3, *
C 20.) FRAGMENTATION 4: (COLUMN 44) THIS VARIABLE IS USED ONLY WITH
C TEETH, MANDIBLES, & VERTEBRA. THE DORSAL/VENTRAL RELATION IS
C CODED HERE.
C
C FOR MANDIBLES:
C INDETERMINATE: 1, INDT           VENTRAL:         4, VTRL
C COMPLETE:      2, COMP           CENTRAL:         5, CENT
C DORSAL:        3, DRSL
C
C FOR TEETH:
C INDETERMINATE: 1, INDT           ROOT:            4, *
C COMPLETE:      2, COMP           CENTRAL:         5, CENT
C CROWN:         3, *
C
C FOR VERTEBRA:
C INDETERMINATE: 1, INDT           ARCH:            3, *
C COMPLETE:      2, COMP           CENTRUM:         4, *
C 21.) TOOTH TYPE: (COLUMN 45) THIS VARIABLE IS USED TO DESCRIBE
C THE TYPE OF BOTH ASSOCIATED AND UNASSOCIATED TEETH.
C
C INDETERMINATE: 1, INDT           PERMENANT:       3, PERM
C DECIDUOUS:     2, DECD
C
C 22.) TOOTH POSITION: (COLUMN 46) THIS VARIABLE DESCRIBES THE
C POSITION OF ASSOCIATED AND UNASSOCIATED TEETH.
C
C INDETERMINATE: 1, INDT           LOWER:           3, LOWR
C UPPER:         2, UPER
C
C 23.) TOOTH CLASS: (COLUMN 47) THIS VARIABLE DESCRIBES THE CLASS
C OF ASSOCIATED AND UNASSOCIATED TEETH.
C
C INDETERMINATE: 1, INDT           PRE-MOLAR:       4, PMLR
C INCISOR:       2, INCI           MOLAR:           5, MOLR
C CANINE:        3, CANI           CHEEK TOOTH:     6, CHEK
C
C 24.) TOOTH NUMBER: (COLUMNS 48-49) THIS VARIABLE DESCRIBES THE NUM-
C BER OF ASSOCIATED AND UNASSOCIATED TEETH.
C
C INDETERMINATE:01, INDT           2 OR 3:          07, 2OR3
C 1:             02, 1             3 OR 4:          08, 3OR4
C 2:             03, 2             1, 2, OR 3:      09, 12R3
C 3:             04, 3             2, 3, OR 4:      10, 23R4
C 4:             05, 4
C 1 OR 2:        06, 1OR2
C
C 25.) TOOTH WEAR: (COLUMNS 50-51) THIS VARIABLE DESCRIBES THE STATE OF
C WEAR OF TEETH OF THE FAMILY BOVIDAE. A SYSTEM DESCRIBING 26 SEPAR-
C ATE STATES OF WEAR DEVISED BY SEBASTIAN PAYNE IS USED AT MICHIGAN.
C THE SYSTEM IS ONLY TO BE USED FOR BOVID TEETH AND IS ESPECIALLY
C DESIGNED FOR OVIS AND CAPRA MANDIBLES.
C
C 26.) TOOTH ROOTING: (COLUMN 52) THIS VARIABLE DESCRIBES THE STAGE
C OF ROOTING OF THE TEETH OF BOVIDAE. BOTH THIS VARIABLE AND VARIABLE
C 25 HAVE BEEN USEFUL IN CONSTRUCTING AGE CURVES. THEY MAY BE DELETED
C IF DESIRED.
C
C INDETERMINATE:        1, INDT    SEPARATE BUT UNCLOSED: 4, O O
C OPEN:                 2, OPEN    CLOSED:                5, CLSD
C SEPARATING AND OPEN:  3, O--O    ROOTS SWOLLEN:         6, RTSW
C
C 27.) RELATIVE AGE: (COLUMN 53) THIS VARIABLE IS TO BE USED WITH
C TEETH ONLY. IN IT A SUBJECTIVE ESTIMATE OF THE AGE OF THE INDIVI-
C DUAL REPRESENTED BY THE TOOTH IS GIVEN. THIS VARIABLE IS NOT USED
C AT MICHIGAN.
C
```

```
C 28.) BURNING: (COLUMN  54) A SUBJECTIVE CLASSIFICATION OF BURNING
C IS USED. THE VALUES MAY BE CHANGED BUT SHOULD ALWAYS PERMIT REDUC-
C TION TO THREE GENERAL VALUES: BURNT, POSSIBLY BURNT, OR UNBURNT TO
C TO FACILITATE COMPARISONS BETWEEN SITES.
C
C INDETERMINATE:          1, INDT         SLIGHTLY AFFECTED:   6, SLTA
C WHITE:                  2, WHTE         POSSIBLY AFFECTED:   7, PSSA
C CARBONIZED:             3, CARB
C BURNT:                  4, BRNT
C PARTIALLY AFFECTED:     5, PRTA
C
C 29.) DISEASE: (COLUMN 55) VALUES FOR THIS VARIABLE WILL VARY FROM
C SITE TO SITE AND INDIVIDUAL TO INDIVIDUAL. THEY SHOULD PERMIT
C GROUPING INTO THE GROSS VALUES USED HERE.
C
C INDETERMINATE:          1, INDT         POSSIBLY DISEASED: 3, PDIS
C DISEASED:               2, DISE
C
C
C 30.) MODIFICATION: (COLUMN 56) ALL MODIFICATIONS EXCEPT BUTCHERING
C MARKS ARE TO BE RECORDED IN THIS VARIABLE. THE LIST BELOW IS NOT
C COMPLETE AND MAY BE ADDED TO.
C
C INDETERMINATE:          1, INDT         CARNIVORE GNAWED:  6, CRNG
C TOOL:                   2, TOOL         GNAWED:            7, GNWD
C WORKED:                 3, WRKD         POSSIBLY WORKED:   8, PWRK
C BITUMEN:                4, BITU         COMBINATION:       9, COMB
C RODENT GNAWED:          5, RODG
C
C 31.) BUTCHERING MARKS: (COLUMN 57) THIS CATEGORY RECORDS THE
C PRESENCE OR ABSENCE OF BUTCHERING MARKS.
C
C INDETERMINATE:                1, INDT   POSSIBLY PRESENT: 3, PSSP
C BUTCHERING MARKS PRESENT: 2, PRES
C
C 32.) SEX: (COLUMN 58) ELEMENTS INDICATIVE OF SEX ARE RECORDED IN
C THIS VARIABLE. ELEMENTS THAT DO NOT ALLOW DETERMINATION OF SEX ARE
C LEFT UNCODED.
C
C INDETERMINATE: 1, INDT              POSSIBLY MALE:     5, PMLE
C MALE:          2, MALE              POSSIBLY FEMALE:   6, PFML
C FEMALE:        3, FEML              POSSIBLY CASTRATE: 7, PCST
C CASTRATED      4, CAST
C
C 33.) DOMESTICATION: (COLUMN 59) THIS VARIABLE IS TO BE USED ONLY
C WITH ELEMENTS WHICH PROVIDE EVIDENCE OF DOMESTICATION. FOR OTHER
C ELEMENTS IT IS LEFT BLANK.
C
C INDETERMINATE:       1, INDT        WILD:              4, WILD
C DOMESTIC:            2, DOMS        POSSIBLY WILD:     5, PWLD
C POSSIBLY DOMESTIC:   3, PDOM
C
C 34.) MEASURE:  (COLUMN 60) IF MEASUREMENTS OF THE BONE WERE TAKEN
C IT IS INDICATED  HERE.
C
C INDETERMINATE: 1, INDT              YES:               2, YES
C
C 35.) COMMENT: (COLUMN 61) IF A COMMENT ABOUT THE ELEMENT WAS MADE
C ELSEWHERE IT IS INDICATED HERE.
C
C YES:   1, YES
C
C 36.) ASSOCIATION: (COLUMNS 62-66) IF A BONE IS ARTICULATED WITH
C ANOTHER BONE, THE NUMBER OF THE BONE IT IS ASSOCIATED WITH IS RECORD-
C CORDED HERE. THIS VARIABLE IS ESPECIALLY DESIGNED FOR INDICATING
C ASSOCIATION OF TEETH WITHIN JAW BONES. IN THIS CASE, THE
C TEETH ARE REFERENCED BY THE NUMBER OF THE JAW ELEMENT WITH
C WHICH THEY ARE ASSOCIATED.
C
C 37.) WEIGHT: (COLUMNS 67-70) THE WEIGHT OF EACH BONE IN
C 1. GRAMS IS RECORDED HERE. THE TOTAL WEIGHTS BY SORTED UNIT
C IS COMPUTED BY BONESORT II.
C
C     **************INITIALIZATION AND CONTROL COMMANDS*********
C
C THE FOLLOWING IS THE REQUIRED ARRANGEMENT FOR THE PROGRAM, DATA AND
C CONTROL CARDS FOR THE MICHIGAN TERMINAL SYSTEM TO BE RUN FROM
C BATCH. SLIGHT MODIFICATIONS MUST BE MADE IF THE PROGRAM IS TO BE
C RUN FROM FROM A TERMINAL.
C
C     1.) $SIGNON CCID P=X T=X
C     2.) (PASSWORD)
C     3.) $RUN *FTN SPUNCH=-LOAD
C     4.) (PROGRAM CARDS)
C     5.) $RUN -LOAD 3=(FILE IN WHICH THE TRANSLATIONS ARE STORED)
C         4=(FILE IN WHICH THE DATA IS STORED) 5=*SOURCE* (OR FILE IN
C WHICH CONTROL CARDS ARE STORED) 6=*PRINT* (OR FILE INTO WHICH
C SORTED DATA ARE TO BE WRITTEN)
C
C     6.) (THE RUN CONTROL CARD)
C
C         COLUMNS 1-3: NUMBER OF SORTING OPERATIONS (EG. 009)
C
```

```
C    7.) (THE DATA CARDS: IF DATA ARE NOT STORED IN A FILE)
C
C    8.) (THE TITLE CARD)
C
C         COLUMNS 1-2 THE NUMBER OF SORTING CRITERIA IN AN OPERATION UP
C                 TO 37
C         COLUMNS 3-78 ANY 116 CHARACTER TITLE FOR AN OPERATION
C
C    9.) (THE COLUMN OF CRITERIA LOCATOR CARD)
C
C      COLUMNS 1-4 VARIABLE NUMBER OF THE FIRST CRITERION
C      COLUMNS 5-8 VARIABLE NUMBER OF THE SECOND CRITERION
C              ECT. UP TO 37 CRITERIA
C
C    10.) (THE STATE LOCATOR CARD)
C
C        COLUMNS 1-4 THE STATE OF CHARACTERISTIC DESIRED WITHIN THE FIRST
C                VARIABLE TO BE SELECTED
C        COLUMNS 5-8 THE STATE DESIRED WITHIN THE SECOND VARIABLE TO BE
C                SELECTED.
C
C EXAMPLE OF CONTROL CARDS
C 002 (NUMBER OF SORTING OPERATIONS)
C (DATA, IF FROM BATCH)
C 03 ALL OVIS RIGHT FEMURS (NUMBER OF SORTING CRITERIA AND TITLE TO
C    BE PRINTED)
C 000900120013 (VARIABLES SPECIES, ELEMENT, SYMMETRY AS CRITERIA TO
C    BE SORTED)
C 001100860002 (STATES OF VARIABLES ABOVE TO BE SORTED, OVIS, FEMUR,
C AND RIGNHT.
C 04 SQUARE 12 LOT 2 BURNT EQUID BONES
C 0003000400120028 (STRATA 1, STRATA 2, SPECIES, BURNING VARIABLES)
C 0001000200270005 (SQUARE 1, LOT 2, EQUID, BURNT STATES)
C
C
C                                     R.W. REDDING
C                                     M.A. ZEDER
C                                     UNIVERSITY OF MICHIGAN
C                                     J. MCARDLE
C                                     UNIVERSITY OF CHICAGO
C                                     JANUARY 11, 1978
C
C
C    ***************VARIABLES USED IN THE PROGRAM*******************
C
C ICOL=VARIABLE (CATAGORY) NUMBER FROM FILE VERB
C ITRANS=TRANSFER VARAIBLES USED IN READING FILE VERB TO DISTINGUISH
C    BETWEEN VARIABLES THAT ARE TRANSLATED AND THOSE THAT
C    REMAIN IN NUMERIC CODE
C TITL1&TITL2=VARIABLE HEADINGS READ FROM FILE VERB
C VERB=FILE NAME AND ARRAY OF VERBAL RESPONSES FOR THE CODED DATA
C ORIG=ARRAY OF THE CODED DATA USUALLY READ FROM A FILE
C NC=THE NUMBER OF SORTING OPERATIONS
C NO=THE NUMBER OF DATA CARDS HERE SPECIFIED AS UP TO 4000
C    THIS NUMBER MAY BE EXPANDED WITH A SLIGHT CHANGE IN THE
C    DIMENSIONING OF ORIG AND ICODE. ONE MUST BE CAREFUL, HOWEVER,
C    NOT TO EXCEED THE STORAGE CAPACITY OF THE SYSTEM IF ENLARGE-
C    MENT OF THIS VARIABLE IS DESIRED.
C IC=INCREMENT VARIABLE
C CAT=THE NUMBER OF CRITERIA TO BE SORTED
C HEAD=THE TITLE OF THE SORTING OPERATION
C COL=THE VARIABLE NUMBER OF THE CATAGORY TO BE SELECTED
C STATE=THE NUMBER OF THE STATE DESIRED WITH THE CATAGORY
C ICODE=TEMPORARY STORAGE ARRAY OF THE LINE NUMBERS ON THE
C    OF SELECTED ELEMENTS
C
C
C    *********************THE PROGRAM**************************
C
C
C VARIABLES ARE DECLARED AND DIMENSIONED
C
      INTEGER CAT,ORIG(37,4000),HEAD(19),COL(46),STATE(200),
     +ICOL(46),ITRANS(46),TITL1(46),TITL2(46),VERB(37,200),TOTAL,
     +ICODE(4000)
C
C THE FILE CONTAINING THE TRANSLATIONS (VERB) IS READ
C
      DO 10 I=1,37
      READ(3,5) ICOL(I),ITRANS(I),TITL1(I),TITL2(I)
    5 FORMAT (I2,I2,1X,A3,A4)
      IF (ITRANS(I) .LE. 0) GO TO 10
      READ(3,6) (VERB(ICOL(I),J),J=1,200)
    6 FORMAT (20A4)
   10 CONTINUE
C
C CODED DATA ARE READ
C
      NO=4000
      DO 28 I=1,NO
   28 READ(4,27,END=30) (ORIG(J,I),J=1,37)
   27 FORMAT (I3,I5,3I4,2I2,I1,I3,I1,I2,I3,I1,I2,2I1,I2,6I1,2I2,
     +10I1,I5,I4)
   30 NO=I-1
```

```
C
C ALL NON-CODED VARIABLES ARE INCREMENTED SO THAT THEY WILL BE
C PRINTED AS BLANKS
C
      DO 50 I=1,NO
      DO 49 J=6,35
      IF (ORIG(J,I) .GE. 1) GO TO 49
      ORIG(J,I)=200
   49 CONTINUE
   50 CONTINUE
C
C THE NUMBER OF SORTING OPERATIONS IS READ FROM THE CONTROL CARDS
C
      READ(5,35) NC
   35 FORMAT (I3)
C
C THE SORTING OPERATION BEGINS
C
      DO 103 II=1,NC
C
C THE INCREMENT VARIABLE IS ESTABLISHED
C
      IC=1
C
C THE NUMBER OF CRITERIA TO BE SORTED AND THE TITLE OF THE SORTING
C OPERATION ARE BOTH READ
C
      READ(5,36) CAT,HEAD
   36 FORMAT (I2,29A4)
C
C THE PARTICULAR CATEGORIES AND STATES TO BE SORTED ARE READ
C
      READ(5,37) (COL(I),I=1,CAT)
      READ(5,37) (STATE(I),I=1,CAT)
   37 FORMAT (5I4)
C
C THE PROCEDURE FOR SELECTING RELEVANT BONES IS BEGUN. EACH BONE IS
C CONSIDERED SEPARATELY BY CATEGORY. IF THE BONE CONSIDERED DOES NOT
C MEET ALL THE SPECIFIED CRITERIA, THE NEXT BONE IS CONSIDERED. IF
C THE BONE DOES MEET THE SPECIFICATIONS, THE NUMBER OF THE LINE OF
C THE CODED BONE IS STORED IN "ICODE".
C
      DO 102 K=1,NO
      DO 101 M=1,CAT
      IF ((ORIG(COL(M),K)) .NE. (STATE(M))) GO TO 102
  101 CONTINUE
      ICODE(IC)=K
      IC=IC+1
  102 CONTINUE
      IC=IC-1
C
C THE PROCEDURE FOR PRINTING THE SELECTED INFORMATION IS BEGUN BY
C PRINTING THE APPROPRIATE TITLE OF THIS PARTICULAR SORTING OPERATION
C
      WRITE(6,29) HEAD
   29 FORMAT ('0',29A4)
C
C IF THERE ARE NO RELEVANT BONES, THE PROGRAM SKIPS TO THE END
C OF THIS PROCEDURE AND WRITES A STATEMENT SAYING SUCH
C
      IF (IC .EQ. 0) GO TO 100
C
C THE FIRST 14 HEADINGS FOR THE DATA ARE PRINTED
C
      WRITE(6,45) (TITL1(I),TITL2(I),I=1,14)
   45 FORMAT ('0',5X,14(A3,A4,2X))
C
C THE DATA PERTAINING TO THE FIRST 14 CATAGORIES ARE TRANSLATED
C AND PRINTED
C
      DO 55 I=1,IC
   55 WRITE(6,31) I,(ORIG(J,ICODE(I)),J=1,5),
     +(VERB(J,ORIG(J,ICODE(I))),J=6,14)
   31 FORMAT (' ',I4,2X,I3,4X,I5,3(5X,I4),1X,9(5X,A4))
C
C THE NEXT 14 HEADINGS ARE PRINTED
C
      WRITE(6,22) HEAD
   22 FORMAT ('0',2X,29A4,'CONTINUED')
      WRITE(6,45) TITL1(2),TITL2(2),TITL1(12),TITL2(12),
     +(TITL1(I),TITL2(I),I=15,26)
C
C THE DATA PERTAINING TO THE NEXT 12 CATAGORIES ARE TRANSLATED
C AND PRINTED
C
      DO 57 I=1,IC
   57 WRITE(6,33) I,(ORIG(2,ICODE(I))),(VERB(12,ORIG(12,ICODE(I)))),
     +(VERB(J,ORIG(J,ICODE(I))),J=15,26)
   33 FORMAT (' ',I4,2X,I5,4X,13(A4,5X))
```

```
C
C THE LAST 13 HEADINGS ARE PRINTED
C
      WRITE(6,22) HEAD
      WRITE(6,45) TITL1(2),TITL2(2),TITL1(12),TITL2(12),
     +(TITL1(I),TITL2(I),I=27,37)
C
C THE TOTAL WEIGHT OF ELEMENTS IN THIS SORTED UNIT IS CALCULATED
C
      TOTAL=0
      DO 58 I=1,IC
      TOTAL=TOTAL+ORIG(37,ICODE(I))
C
C THE REMAINING DATA ARE TRANSLATED AND PRINTED.
C
   58 WRITE(6,34) I,(ORIG(2,ICODE(I))),(VERB(12,ORIG(12,ICODE(I)))),
     +(VERB(J,ORIG(J,ICODE(I))),J=27,35),(ORIG(J,ICODE(I)),J=36,37)
   34 FORMAT (' ',I4,2X,I5,4X,10(A4,5X),I5,5X,I4)
C
C THE TOTAL WEIGHT IS PRINTED
C
      WRITE(6,13) TOTAL
   13 FORMAT (109X,'TOTAL=',I5)
      GO TO 103
C
C A STATEMENT IS WRITTEN BELOW THE FIRST SET OF HEADINGS IF THERE
C ARE NO ELEMENTS WHICH FIT THE SPECIFICATIONS DESIRED.
C
  100 WRITE(6,39)
   39 FORMAT ('0',20X,'THERE ARE NO ELEMENTS WHICH FIT THESE
     +SPECIFICATIONS IN THIS UNIT')
C
C THIS SORTING OPERATION IS COMPLETE AND THE NEXT ONE BEGINS
C
  103 CONTINUE
      STOP
      END
C ********************FILE VERB********************************
C
C     VERB CONTAINS THE TRANSLATIONS OF THE CODE USED IN BONESORT II.
C THE VERSION OF VERB PRESENTED HERE IS USED IN THE ANALYSIS OF NEAR
C EASTERN FAUNAL MATERIAL, SPECIFICALLY FOR THE FAUNA FROM TALL-I
C MALYAN. ANY CHANGES IN THE CODE WITHIN PRESENT FIELD LENGTHS MAY BE
C ACCOMOMODATED BY ALTERING THIS FILE. ANY CHANGES IN THE CODE THAT
C INCLUDE EXPANSION OF THE NUMBER OF FIELDS IN ANY VARIABLE, REQUIRE
C ALTERATIONS IN THE INPUT-OUTPUT FORMATS OF BONESORT II.
C     THE FILE CONSISTS OF VARIABLE TITLE CARDS AND STATE VERBALIZA-
C TION CARDS. THE FIRST 2 COLUMNS OF EACH VARIABLE TITLE CARD CONTAIN
C THE REFERENCE NUMBER OF EACH VARIABLE (01-37). THE 3RD AND 4TH
C COLUMNS CONTAIN EITHER -1 OR (BLANK)1. -1 INDICATES THAT THE NUMBER
C CODED AS INPUT IS TO BE PRINTED, AND IS USED FOR SUCH VARIABLES AS
C BONE NUMBER AND WEIGHT. 1 INDICATES THAT THE NUMERIC INPUT IS TO BE
C TRANSLATED INTO ALPHABETICS, AND IS USED FOR SUCH VARIABLES AS
C PERIOD, CONTEXT, AND SPECIES. COLUMNS 6-12 CONTAIN THE TITLE OF THE
C VARIABLE TO BE PRINTED AS HEADINGS IN BONESORT II OUTPUT. FOLLOWING
C VARIABLE TITLE CARDS OF VARIABLES THAT ARE TO BE TRANSLATED, ARE 10
C STATE VERBALIZATION CARDS. EACH CARD CONSISTS OF 20 FOUR LETTER
C ABBREVIATIONS EACH CORRESPONDING TO A PARTICULAR STATE OF THE VAR-
C IABLE NUMERICALLY CODED IN THE INPUT.  UP TO 200 STATES ARE POSSIBLE
C WITHOUT ANY ALTERATIONS IN THE SORT PROGRAM. ALL 10 CARDS MUST BE
C PRESENT EVEN IF THERE ARE FEWER THAN 200 STATES FOR A VARIABLE.
01-1 SITE
02-1 BONE NO
03-1 STRATA1
04-1 STRATA2
05-1 STRATA3
06 1 PERIOD
INDTMUSHJARIBKUNLAPUBNSHKAFTQLEHMELMLELMPRSAISLM




07 1 CONTEXT
INDTFLORFEATPIT DUMPFILL
```

08 1 CLASS
INDTINVTCHONOSTEAMPHREPTAVESMAMM

09 1 IDENT.
INDTLMAMMMAMSMAMLARTMARTSARTBOVDOCG OVCPOVISCPRAGZLAGDORGGAZGSUBBOS BUBUCERVCPRL
CRVSDAMASUIDSSCRCMLDCMLSEQUDEQAHEQASEQEQEQHMLCRNMCRNSCRNVIVRHRPSHAURHICHHEDWICAL
GEGEHYNALFELMFELSFELACIXCARAFCATFCHAFLEOFMARFPARFUNCLYNXMMUSSMUSLTRAMFOIMEMEMCAP
MAIVVPERLCANMCANSCANCNISCAURCFAMCLUPFENNVLPSVCANVRUPVVULURUSPROCERNDERNCHMIEPRCH
TLPASORICROCNEOMSUNCCHIRRODTSCUICRICCALOCRCTMESOMICRARVIELLOMCRTGRBLGRBSMRIOPSMY
SEKETATRMURDACOMAPODARVCMUS NESORATTDIPOALLAJACUGLIRDRYOGLISSPALHYSTLAGOLEPUOCHO

10 1 RELIABL
INDTRELIPRELNREL

11 1 ECOLOGY

12 1 ELEMENT
INDTBASOOCIPBASPMESEPRSPALISPTERVOMRPALATURBSUPRINTPPARIFRONPSTPSQTPMLARLACRNASL
PMWTPMOTMXWTMXOTMNWTMNOTTOIATONAHORNANTLSTYLEPIHCERABASHTHYHATLSAXISCVRTTVRTLVRT
SACRCAUDRIB CRIBPRSTMSSTXIPHSCAPCLAVHUMRRADIULNARCRPICRPUCRPCCRPPISI1CRP2CRP3CRP
ISPUILLIISHIPUBSOSPNFEMRPATLTBIAFIBULTMLASTRCALCCTRS1TRS2TRS3TRS45TRC4TRMTT1MTT2
45CRRICR23CRMTC1MTC2MTC3MTC4MTC5MC34MTCRPSEFDSEFSESF1PHF2PHF3PHFPHLFINNOILISILPU
MTT3MTT4MTT5MT34MTTRPSEHDSEHSESH1PHH2PHH3PHHPHLHMP34MTPDPRSEDSSESEAS1PHL2PHL3PHL
PHLX2&3T

13 1 SYMETRY
INDTRGHTLEFTMEDL

14 1 FUSION
  INDTFUSDFUSIUFSDPFDFPFDIPFDUPIDUPIDIPIDUPUDFPUDIPUDUFOET

15 1 DEGREEF
INDTCOMPC3/434121214L1/4

```
16 1 ORIGIN
INDTPRED?PREPOST?PSTRECT?RCTCOMB




17 1 FRAG 1
INDTCOMPPENDPSFTPE&SSHFTDE&SDSFTDENDSY&DSYMPDIASCHEK




18 1 FRAG 2
INDTCOMPANT POSTCENT




19 1 FRAG 3
INDTCOMPLAT MEDICENT




20 1 FRAG 4
INDTCOMPDRSLVENTCENT




21 1 TOOTHT
INDTDECDPERM




22 1 TOOTHP
INDTUPPRLOWR




23 1 TOOTHC
INDTINCICANIPMLRMOLRCHEK
```

24 1 TOOTHN
INDT   1   2   3   4   51OR22OR33OR412R323R4

25 1 TOOTHW
INDT   1   2   3   4   5   6   7   8   9 10 11 12 13 14 15 16 17 18 19
       20 21 22 23 24 25 26

26 1 TOOTHR
INDTOPENO--OO OCLSD

27 1 RELAGE

28 1 BURNG
INDTWHTECARBBRNTPRTASLTAPSSA

29 1 DISEASE
INDTDISEPDIS

30 1 MODIFD
INDTTOOLWRKDBITURODGCARGGNWDPWRKCOMB

31 1 BUTCH
INDTPRESPSSP

32 1 SEX
INDTMALEFMALCASTPMLEPFMLPCST

33 1 DOMES
INDTDOMSPDOMWILDPWLD

34 1 MEASURE
INDTYES

35 1 COMMENT
YES

36-1 ASSOC.
37-1 WEIGHT

THE "KNOCOD" SYSTEM FOR PROCESSING DATA ON ANIMAL BONES FROM ARCHAEOLOGICAL SITES

Hans-Peter Uerpmann, Institut für Urgeschichte
University of Tübingen

Introduction

The KNOCOD system for coding and processing data obtained from animal bones recovered from archaeological contexts was developed in 1971 and 1972. Some proposals submitted by Charles A. Reed to the 1971 International Archaeozoological Conference in Budapest provided a starting point. Because several requirements of the author were not met by that system, it had to be modified until now there is left only the basic idea of numerical coding which connects KNOCOD to its intellectual ancestor.

Since the time of its first establishment and use, the KNOCOD system has undergone some changes. The first version of the code was modified in 1973, thus creating a 2nd edition which is still used by the author and some other zooarchaeologists who have adopted this system during the last four years. This 2nd edition contains some inconsistencies because it was designed to make transfer of coded data from the first to the second version as easy as possible. For purposes of publication and further use of KNOCOD, these inconsistencies have now been eliminated; therefore the version published here represents the 3rd edition of this code.

Some basic features of KNOCOD derive from the fact that standard 80-column punch-cards were chosen for data storage because their use facilitates visual control, correction, and hand sorting. The basic unit of information to be stored separately is the set of observations made for one bone specimen. In order to avoid redundant coding as well as to eliminate the danger of losing the connection between two or more punch-cards belonging to the same unit, KNOCOD is designed to accommodate all the usual information from one find on one punch-card. As a result, dense coding is inevitable. For exceptional cases where additional or multiple observations can be made, the system also permits several punch-cards to be joined to form one unit of information. The connection between cards is established by repeating part of the information and by labeling the individual specimens to which the set of data belongs. For final processing, data is usually transferred from punch-cards to more convenient devices such as magnetic tape or disc.

Several programs exist to process KNOCOD data. Except for one elementary sort and select program, all the more refined routines depend upon subroutine packages specific to the computer center used by the author. Therefore no programs are published with this report. A basic feature of most of the programs is the fact that only the input format and not the meaning of the codifications is included in the programs. Thus, for example, the first three columns of a punch-card will be interpreted by the programs as containing information about species without regard to the meaning of any individual code number punched there. Decodification is undertaken only as a final step when generating the output. At this time, code numbers are replaced by texts stored on separate data files. The result is that changes or additions to the numerical code do not require changes in the programs, and except for altering the arrangement of columns, the KNOCOD system provides a high degree of flexibility in permitting alterations to the meanings of the code numbers.

General Description

The information on animal bones handled with the KNOCOD system can be divided into four sections:
1. General information
2. Locational and retrieval information
3. Measurement information
4. Particular information

Referred to as general information are those characters that usually can be observed on an animal bone find: animal species, skeletal element, portion of this element, whether it belongs to the right or left side of the body and to a female or male animal. Frequently found alterations such as burning, butchering marks, and pathologies are also included in this section along with estimates of age at death together with the objective characters of age determination such as tooth eruption and epiphyseal union. Although metrical in nature, the weight of a bone find is also included in the general information category as well as the number of specimens to which the particular unit of information refers.

Locational information includes site and unit of

excavation from where the bone find(s) come(s). The number or other label of the individual specimen, and where appropriate, a code number for the individual animal to which the bone belongs, serve as retrieval information.

Since they are already numerical in nature, measurements require no coding. Every unit of data provides eight positions each containing one measurement. The significance of these positions differs depending upon the codes for species and skeletal element.

For particular information about single specimens, i.e., peculiarities which do not occur often enough to have their own numerical code within the general information section, a short text-field is provided. This field can also be used for any kind of alphanumerical coding of additional information or for amplification of one or more parts of the general, locational, or retrieval information sections.

Description of the Code

Whereas some parts of the KNOCOD code require no comment, most parts need further explanation due to the density of coding mentioned previously. The descriptions and explanations below follow the order of the input format.

Position 1 (Columns 1-3) *Species*: The species code given below is an example of an abridged version developed for use in European and Western Asiatic zooarchaeology. Species which occur only rarely either do not appear in this code or are lumped together under more general groupings. For this code a special translating program exists which groups code numbers into special categories: 001-009 as unidentified, 010-029 as domestic animals, 030-049 as wild or domestic animals, and 050-099 as wild animals. Should species codes be developed for other research areas, it is advisable to maintain the above divisions. In areas where no domestic animals occur, the code numbers 010-049 should not be used.

Position 2 (Columns 4-5) *Skeletal element*: The code for skeletal elements requires little comment. This code serves as the base for the following code (part of bone) and for positioning the measurements. These features explain the double appearance of some elements in the code (see below).

Position 3 (Column 6) *Part of bone*: Although it is understood that a description of bone fragmentation with only nine possible states is not very satisfactory, this seems to be sufficient for a general statement in most cases. To deal with complexities presented by skulls, mandibles, and pelvic bones, however, additional possibilities are necessary. Rather than use up an additional column of the punch-card for these special cases, it was decided to subdivide these items at the level of the element code (position 2). Therefore the code number for skeletal element in these cases depends upon the part of the bone present. If, for instance, a mandible is present in the form of a fragment of the diastema, it is coded as skeletal element no. 11. If the corpus and the ramus are preserved, it is coded as no. 12. The code for the part of bone present is no. 2 in both cases.

Positions 4, 5, and 6 (Columns 7, 8, and 9-10): No comment is necessary on the codes for right or left side of the body, for female or male, or for remarks on special features of individual specimens. These fields are left blank when no information of these kinds can be recorded.

Position 7 (Columns 11-13) *Ageing*: Age information is divided into two sections: first, an estimation of the stage of development reached by the animal at the time of its death is coded; then, objective criteria for ageing are noted. An estimate of the stage of development reached by an animal can often be made on the basis of the size and structure of individual bones. No general rules can be given to define the stages "foetal," "infantile," or "juvenile" when dealing with bones. Decisions must be based on experience and on specific observations for the particular corpus. Bones are classified as "subadult" when structure and size do not suggest that they came from a young animal but ontogenetic processes such as tooth eruption or epiphyseal union have not yet been completed. The classification "adult" is applied in cases when late erupting teeth can be observed to be worn or where late fusing epiphyses are closed. Since many parts of the skeleton do not contain teeth or have late fusing epiphyses, the classification used will often be "subadult or adult" (code no. 8) determined on the basis of structure and size.

Since epiphyseal union and the eruption and wear of teeth can never be observed together on one specimen, they are both coded in the same columns of the punch-card. All combinations of the numbers 0, 1, 2, and 3 are reserved for the code on epiphyseal union (1 = unfused, 2 = in fusion, 3 = fused, 0 = no information). The first of the two columns contains this information for the proximal (or, where appropriate, only) epiphysis of a bone, the second column for the distal epiphysis. Thus the code numbers 01, 02, 03, 10, 11, 12, 13, 20, 21, 22, 23, 30, 31, 32, and 33 are used to specify the stages of epiphyseal union. The remaining code numbers up to 99 are used for coding tooth eruption, replacement, and wear. The code given for wear stages must be precisely defined on the basis of the species and specific wear patterns observed in the particular bone complex under study.

Position 8 (Columns 14-15) *Individual*: In some cases whole skeletons or larger parts thereof are found articulated, or sometimes various bones can be determined to have come from the same individual. Since the set of data from a single bone forms the unit of information (and not the set of data from a single animal) it is necessary to combine these units by labeling bones from one individual with the same number. Different individuals of the same species within one site or within one unit of excavation (where appropriate) must get different labels. For unconnected bone finds this position is left blank except in the case where one bone is represented by more than one set of data. This situation may occur when age information on more than one tooth of the same jaw is to be recorded or when the measurements will not fit onto only one punch-card (see below). In this case, species information and the appropriate skeletal element code are repeated as are retrieval and locational information together with those specifications which would not fit onto the previous punch-card(s). The connection between multiple data sets of this kind is also marked by giving the same individual label to each of the cards included in the set.

Position 9 (Columns 16-18) *Weight*: The weight of the bone find(s) dealt with on the respective card is entered right-justified in columns 16-18. The existing programs expect the weight to be given in grams. Weights must not be repeated when multiple punch-cards are coded for one find.

Position 10 (Columns 19-20) *Number of pieces*: The number of bone finds to which one set of data applies will usually be one (1). In the case of unidentified fragments or ribs, however, several bones may be characterized by identical information. The number of pieces with identical information is entered right-justified in columns 19-20. Where multiple punch-cards are to be coded for one specimen, this field must be left blank or filled with zeros on all but the first card in order to avoid multiple counting during quantitative analysis. For single specimens the number 1 is coded in column 20.

Position 11 (Columns 21-24) *Label*: The unique label or number of a bone specimen can be entered in this field to serve as a means of retrieval. Where bones do not have individual labels it is possible instead to use this field for uniquely labeling individual data units.

Position 12 (Columns 25-30) *Excavation unit*: Because of the considerable variation in excavation techniques and recording systems, no general rules for codification of this position can be given here. Where it is impossible to enter without change the designations assigned by the excavator into this position, individual codes must be developed. If more than six columns are needed, part of the information can be entered in the text field at the end of the data line.

Position 13 (Columns 31-34) *Site*: Just as with units of excavation, no general rules can be given for the coding of sites. The author uses a code based on a rough geographical ordering.

Position 14 (Columns 35-36) *Measurement characterization*: This position marks the beginning of the measurement information fields and defines the quality of the measurements. Entries in column 35 are also used to simplify processing by indicating whether in fact there exist measurements for the particular bone.

Measurements (Columns 37-68): There are eight positions of four columns each available on each punch-card to enter measurements. These are taken to the nearest tenth or half millimeter and are entered right-justified without any decimal point. The significance of each position depends upon the skeletal element being measured and, to a certain extent, on the animal species concerned. Measurement position one, for example, may contain the length of the cheek tooth row when dealing with a mandible, or the distal breadth when dealing with a humerus. A detailed table of how to enter those measurements defined by von den Driesch (1976) is given below (Table 2). All measurement abbreviations used in that table as well as all figures referred to are found in that handbook.

Special problems arise with skulls and mandibles because more than eight measurements often must be taken. Because these bones are subdivided already within the element code in order to provide sufficient possibilities for describing fragmentation, these subdivisions are used to enter additional measurements. In cases where more or different measurements are to be recorded, additional punch-cards must be prepared by repeating species, retrieval, and locational information (as noted above under Position 8) and by adjusting the element code according to what measurements are to be taken. Number and weight must be entered as zero in the appropriate places on these additional data cards; only on that card where the specimen is completely described by the element and fragmentation code are values other than zero entered for weight and number. An example of how to code information on a complete skull of a dog missing its teeth is given in Table 1. This example also demonstrates the use of Position 14. In the first four data lines, the code number 5 indicates that normal measurements follow. In the last line, the code number 1 indicates that the measurement specified in the next column (= 4) is only an estimate (in this example the length of the P^4 estimated from the alveolar measurement).

In a similar fashion, complete mandibles must be coded using element codes 11 (tooth row measurements, etc.), 12 (corpus length measurements), and 13 (or 92 and 94 - measurements of single teeth). The measurements of all other parts of the skeleton can be recorded on a single card, although for special purposes usage similar to that for the skull is possible for other elements.

Comment (Columns 69-80): No explanation is necessary for the use of the text field available at the end of every data line (see general description above). For those who used the first or second version of the KNOCOD code, it is necessary to punch the number of the code edition used in the last column of all data cards to prevent misinterpretations.

TABLE 1. Coding for a complete dog skull without teeth

Position 1	2	3	4	5	6	7	8	9	10	11	12/13	14	M1	M2	M3	M4	M5...
015	09	4	5	3		882	1	295	1	HU03	(loc.-sit)	5	2110	2265	2000		......
015	03					8		1	0	0HU03	(loc.-sit)	5	465	445	700	560	......
015	04					8		1	0	0HU03	(loc.-sit)	5	1030	740	685		......
015	07					8		1	0	0HU03	(loc.-sit)	5	715	175	590		......
015	08					8		1	0	0HU03	(loc.-sit)	14				190	......

Format for Input (on 80-column punch-card)

Column	Position	Description
1		
2	1	Species
3		
4	2	Skeletal element
5		
6	3	Part of bone
7	4	Side
8	5	Sex
9		
10	6	Remarks
11	7.1	Routh ageing
12	7.2	Epiphyseal fusion
13	7.3	Tooth eruption and wear
14	8	Individual
15		
16		
17	9	Weight
18		
19	10	Number of pieces
20		
21		
22	11	Label or number of bone
23		
24		
25		
26		
27	12	Excavation unit
28		
29		
30		
31		
32		
33	13	Site
34		
35	14	Measurement characterization
36		
37		
38	M1	Measurement 1
39		
40		
41		
42	M2	Measurement 2
43		
44		
45		
46	M3	Measurement 3
47		
48		

Column	Position	Description
49		
50	M4	Measurement 4
51		
52		
53		
54	M5	Measurement 5
55		
56		
57		
58	M6	Measurement 6
59		
60		
61		
62	M7	Measurement 7
63		
64		
65		
66	M8	Measurement 8
67		
68		
69		
70		
71		
72		
73	CM	Comment
74		
75		
76		
77		
78		
79		
80	CE	Code edition (=3)

Input Codes

1. <u>Species</u> (Columns 1-3)

1.1 Unidentifiable
   001 no size assignment
   002 smaller than rabbit-size
   003 rabbit-size to medium size dog ("<u>Canis pallustris</u>")
   004 medium size dog to medium size sheep (roe deer, gazelle)
   005 medium size dog to wild boar-size (ibex, fallow deer)
   006 Wild boar-size to red deer, small cattle-size
   007 medium size sheep to domestic cattle-size
   008 European red deer-size to domestic cattle-size
   009 larger than domestic cattle-size

1.2 Domestic animals of lesser economic
         importance
    010 domestic hen
    011 other domestic fowl
    012 domestic rabbit
    013
    014 domestic cat
    015 dog
    016
    017 horse
    018 donkey
    019 mule

1.3 Domestic animals of greater economic
         importance
    020 pig
    021 sheep
    022 camel
    023 goat
    024
    025 sheep or goat
    026 cattle
    027
    028
    029

1.4 Wild or domestic form of first group
    030 wild or domestic hen
    031 wild or domestic fowl
    032 wild or domestic rabbit
    033 rabbit or hare
    034 wild or domestic cat
    035 wolf or dog
    036 jackal or dog OR fox or dog
    037 wild or domestic horse
    038 wild or domestic donkey
    039 equid

1.5 Wild or domestic form of second group
    040 wild or domestic pig
    041 wild or domestic sheep
    042 wild or domestic camel
    043 wild or domestic goat
    044
    045 wild or domestic sheep or goat
    046 wild or domestic cattle
    047 large cervid or bovid
    048 small cervid or ovi-caprid
    049 small artiodactyl

1.6 Wild form of first group
    050 wild hen
    051 wild ducks or geese
    052 wild rabbit
    053 wild rabbit or hare
    054 wild cat
    055 wolf
    056 jackal
    057 wild horse
    058 wild ass
    059 wild equid

1.7 Wild form of second group
    060 wild boar
    061 wild sheep
    062 wild camel
    063 wild goat
    064 ibex
    065 wild goat or sheep
    066 aurochs
    067 bison
    068 other large bovid
    069

1.8 Hunted wild animals - artiodactyls
    070 red deer
    071 fallow deer
    072 elk (moose)
    073 roe deer
    074 not identifiable or other cervid
    075 reindeer
    076 musk ox
    077 gazelle
    078 chamois
    079 other wild ruminant

1.9 Wild animals - carnivores
    080 bear
    081 badger
    082 fox
    083 lynx
    084 not identifiable or other carnivore
             of medium size
    085 martens
    086 polecats
    087 weasels
    088 not identifiable or other small
             carnivore
    089 not identifiable or other large
             carnivore

1.10 Wild animals - hares, rodents, insectivores,
         non-mammals
    090 hare
    091 beaver
    092 not identifiable or other large rodent
    093 small rodent
    094 hedgehog
    095 insectivores and bats
    096 birds except domesticated or major
             food species
    097 reptiles
    098 amphibians
    099 fish

2.   Skeleton (Columns 4-5)
    01 horn core or antler
    02 part of skull with horn core or antler
    03 skull fragment, brain case only
            (PART 1; Sec. 3.4.1)
    04 skull fragment, brain case only
            (PART 2; Sec. 3.4.2)
    05 brain case and face fragment
    06 face bones, except mandible
            (PART 1; Sec. 3.4.4)
    07 face bones, except mandible
            (PART 2; Sec. 3.4.5)
    08 loose maxillary tooth
    09 well-preserved skull
    10
    11 fragment of mandible
            (PART 1; Sec. 3.5.1)
    12 fragment of mandible
            (PART 2; Sec. 3.5.2)

2. Skeleton (continued)
   - 13 loose mandibular tooth
   - 14
   - 15 maxillary or mandibular tooth
   - 16 hyoid
   - 17
   - 18
   - 19
   - 20
   - 21 clavicle
   - 22 caracoid
   - 23 scapula
   - 24 humerus
   - 25 radius
   - 26 ulna
   - 27 radius and ulna
   - 28 carpal
   - 29
   - 30 metacarpus III+IV (artiodactyls)
   - 31 metacarpus III
   - 32 metacarpus IV
   - 33 other metacarpus
   - 34 indeterminate metacarpus
   - 35 phalanx 1, anterior
   - 36 phalanx 2, anterior
   - 37 phalanx 3, anterior
   - 38
   - 39
   - 40 fragment of pelvis (PART 1; Sec. 3.8.1)
   - 41 fragment of pelvis (PART 2; Sec. 3.8.2)
   - 42 femur
   - 43 patella
   - 44 tibia
   - 45 fibula
   - 46 astragalus
   - 47 calcaneus
   - 48
   - 49 tarsal (remaining)
   - 50 metatarsus III+IV (artiodactyls)
   - 51 metatarsus III
   - 52 metatarsus IV
   - 53 other metatarsus
   - 54 indeterminate metatarsus
   - 55 phalanx 1, posterior
   - 56 phalanx 2, posterior
   - 57 phalanx 3, posterior
   - 58
   - 59
   - 60 indeterminate metapodial of main axis
   - 61 indeterminate metapodial - peripheral
   - 62 indeterminate metapodial
   - 63 sesamoid
   - 64
   - 65 phalanx 1, anterior/posterior
   - 66 phalanx 2, anterior/posterior
   - 67 phalanx 3, anterior/posterior
   - 68
   - 69 peripheral phalanx, indeterminate
   - 70 atlas
   - 71 axis
   - 72 cervical vertebra
   - 73 thoracic vertebra
   - 74 lumbar vertebra
   - 75 sacrum
   - 76 caudal vertebra
   - 77
   - 78
   - 79 indeterminate vertebra
   - 80 rib
   - 81
   - 82 sternum
   - 83 cartilage part of scapula
   - 84 cartilage part of rib
   - 85
   - 86
   - 87
   - 88
   - 89
   - 90
   - 91 upper premolar
   - 92 lower premolar (additional code for
   - 93 upper molar   measurements on teeth)
   - 94 lower molar
   - 95
   - 96
   - 97
   - 98
   - 99

3. Part of bone (Column 6)

3.1 horn core (01=corresponding code in position 2)
   - 0 indeterminate fragment
   - 1 base
   - 2 corpus
   - 3 tip
   - 4 base + corpus
   - 5 corpus + tip
   - 6
   - 7
   - 8 almost complete
   - 9 complete

3.2 antler (01)
   - 0 indeterminate fragment
   - 1 base to beginning of the first tine
   - 2 first tine
   - 3 base + first tine
   - 4 fragment of column
   - 5 column with tines
   - 6 crown
   - 7 single tine or fragments
   - 8 base, first tine, and column with tines
   - 9 almost complete or complete branch

3.3 horn core or antler with skull (02)
   - 0 indeterminate fragment from transition between frontal and core
   - 1 frontal with one horn base
   - 2 frontal with almost complete core
   - 3 frontal with complete core
   - 4 frontal with other parts of skull and measurable base or core
   - 5 frontal with two horn core bases
   - 6 frontal with one core base and one complete core
   - 7 frontal with two almost complete or complete cores
   - 8 frontal with more parts of skull + two measurable bases
   - 9 frontal with more parts of skull + two complete cores

3.4.1 braincase 1 (03)
   0 more or less complete braincase
   1 frontal fragment
   2 fragment of frontal and parietal
   3 fragment of parietal
   4 fragment of parietal and squamosal
   5 fragment of squamosal
   6 fragment of squamosal and basioccipital
   7 fragment of basioccipital
   8 fragment of basioccipital and base of cranium
   9 fragment of 6+3 or 6+2 with or without temporal

3.4.2 braincase 2 (04)
   0 other or unidentified fragment of braincase
   1 frontal, parietal, squamosal temporal
   2 frontal, parietal, squamosal
   3 parietal, squamosal, temporal
   4 parietal, temporal
   5 frontal, temporal
   6 squamosal, temporal
   7 temporal
   8 temporal and base of cranium
   9 fragment of base (basal +/- pterygoid)

3.4.3 braincase and face (05)
   0 unidentified braincase fragment +/- face fragment
   1 ethmoid
   2 frontal, nasal
   3 frontal, nasal, lacrimal
   4 frontal, lacrimal
   5 frontal, lacrimal, zygomatic
   6 temporal, zygomatic
   7 fragment from region where nasal opens nuchally and basally)
   8 large part of braincase with part of face
   9 large part of face with part of braincase

3.4.4 face 1 (06)
   0 unidentifiable or other
   1 zygomatic
   2 lacrimal
   3 zygomatic, lacrimal
   4 nasal
   5 premaxillary
   6 maxillary - outer part without alveolar region
   7 maxillary - inner part without alveolar region or palatine
   8 vomer
   9 interior nasal bones

3.4.5 face 2 (07)
   0 almost complete or complete face
   1 maxillary - outer part with alveolar region
   2 maxillary - inner part with alveolar region
   3 maxillary - inner and outer parts with alveolar
   4 maxillary - (outer part +/- inner part) and premaxillary
   5 maxillary - (outer part +/- inner part) and palatine
   6 maxillary - (outer part or inner + outer parts) and nasal
   7 maxillary - (outer part or inner + outer parts) and lacrimal
   8 maxillary - (outer part or inner + outer parts) and zygomatic
   9 maxillary - (outer part or inner + outer parts) and at least two other face bones

3.4.6 complete skull (09)
   0 more or less complete skull
   1 complete skull being damaged on one side
   2 complete half of skull (sagittal section)
   3 complete skull with anterior portion cut off
   4 complete skull with missing teeth
   5 complete skull with missing base
   6 complete skull with missing occipital
   7 complete skull with missing roof
   8 complete skull with missing mouth area and base
   9 complete skull with missing mouth area and base and roof

3.5.1 mandible 1 (11)
   0 unidentified
   1 incisor area + symphysis
   2 diastema
   3 incisor area, symphysis, diastema
   4 corpus - alveolar region
   5 corpus - basal region
   6 corpus - alveolar and basal regions
   7 ramus without processus articularis or muscularis
   8 processus muscularis
   9 processus articularis

3.5.2 mandible 2 (12)
   0 unidentified
   1 corpus, diastema, +/- incisor region
   2 corpus + ramus
   3 corpus, diastema, ramus, +/- incisor region
   4 flat part and at least one process of ramus
   5 upper part of ramus with both processes and small part of flat area
   6 upper part of ramus with both processes and a large part of flat area
   7
   8
   9 complete mandible

3.6 teeth (08, 13, or 15)
   0 indeterminate
   1 milk incisor
   2 incisor
   3 milk premolar
   4 premolar
   5 molar
   6 canine
   7 premolar or molar
   8 third molar fragment (for carnivores, first molar fragment)
   9 complete third molar (for carnivores, first molar or fourth premolar)

156 Approaches to Faunal Analysis

3.7   scapula (23)
- 0
- 1 fragment of flat part
- 2 spine fragment
- 3 fragment of flat part + spine
- 4 articular area
- 5 articular area and part of flat part +/- spine
- 6 isolated thoracic margin
- 7 neck
- 8 almost complete
- 9 complete

3.8.1   pelvis 1 (40) [acetabulum not measurable or missing]
- 0
- 1 flat part of ilium
- 2 column of ilium
- 3 flat part and column of ilium
- 4 acetabulum fragment
- 5 column of ilium with part of acetabulum
- 6 whole ilium with acetabulum fragment
- 7 ischium fragment
- 8 pubis fragment
- 9 ischium and pubis fragment

3.8.2   pelvis 2 (41)
- 0 measurable isolated acetabulum
- 1 ilium or part of ilium and measurable acetabulum
- 2 ischium with measurable acetabulum
- 3 ilium + ischium with measurable acetabulum
- 4 pubis with measurable acetabulum
- 5 pubis + ilium with measurable acetabulum
- 6 pubis + ischium with measurable acetabulum
- 7 pubis + ilium + ischium with measurable acetabulum
- 8 almost complete
- 9 complete

3.9   humerus, radius, ulna, femur, tibia, fibula, metapodia, phalanx 1, phalanx 2 with fused epiphyses
- 0
- 1 proximal end with up to 1/4 of the shaft
- 2 proximal end with more than 1/4 of the shaft
- 3 shaft (no articular ends)
- 4 distal end with more than 1/4 of the shaft
- 5 distal end with up to 1/4 of the shaft
- 6
- 7 bone split longitudinally
- 8 almost complete bone
- 9 complete bone

3.10   same as 3.9 but with one or more epiphyses unfused
- 0 loose proximal epiphysis
- 1 proximal part of diaphysis without epiphysis
- 2 proximal part of diaphysis and loose epiphysis
- 3 large part of diaphysis reaching one open fusion point
- 4 distal part of diaphysis without epiphysis
- 5 loose distal epiphysis
- 6 distal part of diaphysis and loose epiphysis
- 7 complete or almost complete diaphysis without epiphyses
- 8 complete or almost complete diaphysis with one epiphysis
- 9 complete or almost complete diaphysis with all epiphyses

3.11   phalanx 3, sesamoid, patella
- 0
- 1 fragment
- 2
- 3
- 4
- 5
- 6
- 7
- 8 almost complete
- 9 complete

3.12   carpal (28)
- 0 complete but not further identified
- 1 fragment - not identified
- 2 complete radial carpal
- 3 fragment of radial carpal
- 4 complete intermediate carpal
- 5 fragment of intermediate carpal
- 6 complete ulnar carpal
- 7 fragment of ulnar carpal
- 8 complete accessory carpal
- 9 fragment of accessory carpal

3.13   astragalus (46)
- 0
- 1 lateral half
- 2 medial half
- 3 proximal half
- 4 distal half
- 5 small fragment
- 6 large fragment
- 7
- 8 almost complete
- 9 complete

3.14   calcaneum (47)
- 0 proximal extremity or loose epiphysis
- 1 body of tuber
- 2 body of tuber and proximal extremity
- 3 articular part without distal point
- 4 distal point
- 5 3 + 1
- 6 3 + 2
- 7 3 + 4
- 8 3 + 4 + 1
- 9 complete calcaneum

3.15   tarsal (49)
- 0 complete but not further identified
- 1 fragment - not identified
- 2 centrotarsale
- 3 fragment of centrotarsale
- 4 tarsale centrale
- 5 fragment of tarsale centrale
- 6 tarsale tertium
- 7 fragment of tarsale tertium
- 8 os tarsale
- 9 fragment of os tarsale

3.16   atlas (70)
       0
       1   ventral arch
       2   dorsal arch
       3   ventral and dorsal arches without
           wings
       4   wing
       5   dorsal arch and wing
       6   fragment of articular region
       7   ventral and dorsal arches and
           wing(s)
       8   large fragment
       9   complete or almost complete atlas

3.17   axis (71)
       0   dens and cranial part of articular
           surface
       1   body
       2   arch and articular process
       3   body and arch and articular process
       4   transverse process
       5   dens and cranial part of articular
           surface and body
       6   spinous process
       7   split axis
       8   large fragment
       9   complete or almost complete axis

3.18   sacrum (75)
       0
       1   corpus
       2   arch and articular process
       3   corpus and arch and articular process
       4   wing fragment
       5   large fragment of anterior portion
       6   single segment
       7   split sacrum
       8   complete anterior portion
       9   complete or almost complete sacrum

3.19   vertebrae [except atlas, axis, sacrum]
       0
       1   body
       2   arch and articular process
       3   body and arch and articular process
       4   transverse process
       5   arch and articular process and trans-
           verse process
       6   spinous process
       7   split vertebra
       8   arch and articular process and spinous
           process
       9   complete or almost complete vertebra

3.20   rib (80)
       0
       1   head to angle
       2   corpus
       3   head, angle, and corpus
       4
       5
       6
       7
       8   almost complete rib
       9   complete rib

4.   Side [right-left] (Column 7)
       0   not determined
       1   right
       2   left
       3   probably right
       4   probably left
       5   left and right
       6   right medial
       7   right lateral
       8   left medial
       9   left lateral

5.   Sex (Column 8)
       0   not determined
       1   female
       2   male
       3   probably female
       4   probably male
       5   castrate
       6   probably castrate
       7   male or castrate
       8   female or castrate
       9

6.   Remarks (Columns 3-10)
       00   none
       01   comment noted elsewhere
       02
       03   artifact
       04   artifact with comment noted
       05   debitage
       06   debitage with comment noted
       07   gnawing - not further identified
       08   gnawing - carnivore
       09   gnawing - rodent
       10   undetermined burning
       11   trace of cooking
       12   partially charred (except 11)
       13   completely charred
       14   charred, partially calcined
       15   partially calcined
       16   completely calcined
       17
       18
       19
       20   trace of butchering - not further
            defined
       21
       22   [typical traces of butchering to be
            defined for each site]
       23
       24   trace of blow with blunt edge
       25   trace of blow with sharp edge - cut
            through
       26   trace of blow with sharp edge - not
            cut through
       27   slight cut mark
       28   strong cut mark
       29   mark of cut or blow
       30   bone artifact with burning
       31   bone artifact with trace of cooking
       32   bone artifact with partial charring
       33   bone artifact with complete charring
       34   bone artifact with charring and
            partially calcined
       35   bone artifact, partially calcined
       36   bone artifact, completely calcined
       37
       38
       39
       40
       41
       42
       43

6. Remarks (continued)
   44
   45
   46
   47
   48
   49
   50
   51 debitage with trace of cooking
   52 debitage with partial charring
   53 debitage with complete charring
   54 debitage with charring and partly calcined
   55 debitage, partly calcined
   56 debitage, completely calcined
   57
   58
   59
   60 burned with marks of butchering
   61 trace of cooking and marks of butchering
   62 [typical traces of butchering (see 22-23) and burned]
   63
   64 burned and trace of blow with blunt edge
   65 burned and trace of blow with sharp edge - cut through
   66 burned and trace of blow with sharp edge - not cut through
   67 burned and with slight cut mark
   68 burned and with strong cut mark
   69 burned and with mark of cut or blow
   70
   71
   72
   73
   74
   75
   76
   77
   78
   79
   80
   81 preliminary identification
   82 probable identification
   83 possible identification
   84
   85 definite identification
   86
   87
   88
   89
   90 pathological - not otherwise defined
   91 typical alteration for particular part of skeleton
   92 broken without dislocation
   93 broken with dislocation
   94 exostosis at borders of articular surface
   95 exostosis away from articular surface
   96 hyperplasia without inflamation
   97 lesions and grind marks in articulation
   98 lesions away from articulation
   99 pathologically altered and comment noted elsewhere

7. Ageing (Columns 11 and 12-13)

7.1 rough ageing
   0 no age determination
   1 not at all grown up (foetal to juvenile)
   2 foetal or infantile
   3 infantile
   4 infantile or juvenile
   5 juvenile
   6 juvenile or sub-adult
   7 sub-adult
   8 sub-adult or adult
   9 adult

7.2+3 epiphyses and teeth
   00 no age determination
   01 proximal no evidence, distal unfused (UF)
   02 proximal no evidence, distal fusing (EL [=epiphyseal line])
   03 proximal no evidence, distal fused (FU)
   04 premolar not changed
   05 premolar changing
   06 premolar in line but not in wear
   07 premolar slightly worn
   08 premolar moderately worn
   09 premolar heavily worn
   10 proximal UF, distal no evidence
   11 proximal UF, distal UF
   12 proximal UF, distal EL
   13 proximal UF, distal FU
   14 first molar not present
   15 first molar erupting
   16 first molar in place but not in wear
   17 first molar slightly worn
   18 first molar moderately worn
   19 first molar heavily worn
   20 proximal EL, distal no evidence
   21 proximal EL, distal UF
   22 proximal EL, distal EL
   23 proximal EL, distal FU
   24 second molar not present
   25 second molar erupting
   26 second molar in place but not in wear
   27 second molar slightly worn
   28 second molar moderately worn
   29 second molar heavily worn
   30 proximal FU, distal no evidence
   31 proximal FU, distal UF
   32 proximal FU, distal EL
   33 proximal FU, distal FU
   34 third molar not present
   35 third molar erupting
   36 third molar in place but not in wear
   37 third molar slightly worn
   38 third molar moderately worn
   39 third molar heavily worn
   40 first incisor not changed
   41 first incisor changing
   42 first incisor in line but not or only slightly worn
   43 first incisor moderately worn
   44 first incisor heavily worn
   45 second incisor not changed
   46 second incisor changing
   47 second incisor in line but not or only slightly worn
   48 second incisor moderately worn
   49 second incisor heavily worn
   50 third incisor not changed
   51 third incisor changing
   52 third incisor in line but not or only slightly worn
   53 third incisor moderately worn
   54 third incisor heavily worn

55 fourth incisor or canine not changed
56 fourth incisor or canine changing
57 fourth incisor or canine in line but not or slightly worn
58 fourth incisor or canine moderately worn
59 fourth incisor or canine heavily worn
60 milk premolar breaking through
61 milk premolar in line but not worn
62 milk premolar slightly to moderately worn
63 milk premolar moderately to heavily worn
64
65 milk incisor breaking through
66 milk incisor in line but not worn
67 milk incisor slightly to moderately worn
68 milk incisor moderately to heavily worn
69
70 unidentified molar or premolar without developed roots
71 unidentified molar or premolar not worn
72 unidentified molar or premolar slightly worn
73 unidentified molar or premolar moderately worn
74 unidentified molar or premolar heavily worn
75 first incisor changed but not present
76 second incisor changed but not present
77 third incisor changed but not present
78 fourth incisor or canine changed but not present
79 premolars changed but not present
80 first molar erupted but not present
81 second molar erupted but not present
82 third molar erupted but not present
83
84
.
.
.

8. Individual (Columns 14-15)

   right justified number to specify individual animal to which data line belongs

9. Weight of piece(s) (Columns 16-18)

   to the nearest gram; right justified

10. Number of pieces (Columns 19-20)

    right justified

11. Label or Number of specimen (Columns 21-24)

12. Excavation unit (Columns 25-30)

    coded individually for every site according to excavation system

13. Site (Columns 31-34)

14. Measurement characterization (Columns 35-36)

    column 35
    0 no measurement following
    1 measurement 'n' (specified in column 36) is estimated
    2 measurements 'n' and 'n+1' are estimated
    3 measurements 'n' and 'n+2' are estimated
    4 measurements 'n' and 'n+3' are estimated
    5 only normal measurement(s) follow(s)
    6 all measurements except 'n' are estimated
    7 all measurements are estimated
    8 measurement 'n' is influenced by pathology or abnormality
    9 all measurements are influenced by pathology or abnormality

    column 36
    1-8 position 'n' of estimated or influenced measurement

15. Measurements (Columns 37-68)

    Measurements to the nearest tenth or half millimeter are entered right justified in 8 positions consisting of 4 columns each without decimal point. The meaning of each position depends on the skeletal element and to a certain extent on the animal species concerned. The order of entry for the measurements defined by von den Driesch (1976) appears in Table 2.

Acknowledgments

I wish to thank Kathleen Biddick for her help with the English formulation of this text. She also undertook the major task of arranging the measurements defined by von den Driesch (1976) into Table 2. I am also indebted to Richard H. Meadow for a translation of the second version of the code on which this third version is based and for the editing of this paper.

Reference

Driesch, A. von den
 1976 *A Guide to the Measurement of Animal Bones from Archaeological Sites.* Peabody Museum Bulletin 1. Cambridge (Harvard University).

160 Approaches to Faunal Analysis

TABLE 2. KNOCOD - Order of entry for measurements

Position M1 (Columns 37-40)	Position M2 (Columns 41-44)	Position M3 (Columns 45-48)	Position M4 (Columns 49-52)
Horn core (01)			
greatest diameter of horn core base (f.8 #45)	least diameter of horn core base (f.8 #46)	basal circumference of the horn core (Bos #44; f.10 #40)	length of the outer curvature of the horn core (f.8 #47)
Horn cores (pair: 02)			
greatest breadth between the lateral borders of the horn core bases (f.9 #32)	least breadth between the bases of the horn cores (f.8 #31)	greatest tangential distance between the outer curves of the horn cores (f.8 #43)	least distance between the horn core tips (f.8 #42)
Antler (01)			
greatest diameter of burr to be taken at location shown in f.11 #41	greatest diameter of the pedicel to be taken at location shown in f.11 #40	distal circumference of the burr (f.11 #41)	circumference of the distal end of the pedicel (f.11 #40)
Skull, part 1 [ungulates] (03)			
least breadth between the orbitals Entorbitale-Entorbitale (f.5a #42; f.7a #26; f.8a #34; f.9a #35; f.11a #33; Sus as above)	least frontal breadth (f.5a #39; f.7a #24; f.8a #32; f.11a #31; f.18a #15)	least breadth between temporal lines (f.9a #31; f.12c #40)	greatest mastoid breadth (f.5d #33; f.7c #18; f.8d #25; f.9d #26; f.11c #25; f.12e #33)
Skull, part 2 [ungulates] (04)			
length of the frontal along the contour (caprids only)	frontal length (f.8a #8; f.9a #10; f.11a #10; f.12a #14; f.18a #7)	upper neurocranium length (f.5a #9; f.9a #11; f.12a #11)	basicranial length Basion-Synsphenion (for Equus as in f.14c #4)
Maxilla and face [ungulates] (07)			
length of the cheek-tooth row (alveoli) (f.5b #22; f.7c #13a; f.8d #20; f.9c #21; f.11c #20; f.12d #27; f.18b #9)	length of the molar row (alveoli) (f.5c #23; f.7c #14; f.8d #21; f.9c #22; f.11c #21; f.12d #28)	length of the premolar row (alveoli) (f.5c #24; f.7c #15; f.8d #22; f.9c #23; f.11c #22, f.12d #29)	length from $PM^{2(1)}$ to Prosthion
Skull, part 3 [ungulates] (09)			
condylobasal length (f.5b #2; f.7c #2; f.8d #2; f.9c #2; f.11c #2; f.12d #2; f.18a #2)	total length (f.5a #1; f.7a #1; f.8a #1; f.9a #1; f.11a #1; f.12a #1; f.18a #1)	condylobasal length without premaxilla (sheep/goat)	greatest length without premaxilla
Skull [carnivores] (03)			
least breadth between orbits (f.14a #33; f.17a #25)	least breadth of skull (f.14a #31; f.17a #28)	frontal breadth (f.14a #32; f.17a #24)	height of occipital triangle (f.14d #40; f.17b #32)

Position M5 (Columns 53-56)	Position M6 (Columns 57-60)	Position M7 (Columns 61-64)	Position M8 (Columns 65-68)
Horn core (01)			
length of the inner curvature of the horn core	----	----	----
Skull, part 1 [ungulates] (03)			
greatest breadth of occipital condyles (f.5d #34; f.7c #19; f.8b #26; f.9d #27; f.11c #26; f.12e #34)	greatest diameter of foramen magnum	greatest neurocranium breadth Euryon - Euryon (f.5a #38; f.7a #23; f.9a #33; f.11 - not shown)	greatest zygomatic breadth (for <u>Bos</u> as below - f.11c #34; f.12a #43)
Skull, part 2 [ungulates] (04)			
medial length of the parietal (f.9a #9; f.12a #13; f.18a #6)	height of the occipital from the condyles to the occipital/parietal suture	height of the occipital region from the condyles to the highest point of the intercornual ridge (caprids only)	height from the occipital condyles to the Nasion (caprids only)
Maxilla and face [ungulates] (07)			
breadth of the Tubera malaria (f.8a #35; f.9a #36)	greatest length of the nasals (f.5a #16; f.8a #12; f.9a #15; f.11a #15; f.12a #15; f.18a #5)	dental length (f.5c #19; f.7c #11; f.8d #17; f.9c #18; f.12d #18; f.18a #4)	greatest palatal breadth (f.5c #48; f.7c #29; f.8d #38; f.9c #39; f.11c #37; f.12d #44)
Skull, part 3 [ungulates] (09)			
facial length Supraorbitale-Prosthion (f.5a #10; f.9b #12; f.12a #12)	greatest breadth (f.5a #41; f.7a #25; f.8a #33; f.9b #34; f.11a #32; f.12a #43)	Nasion-Prosthion (f.12a #10; and so for other ungulates)	----
Skull [carnivores] (03)			
greatest breadth occipital condyles f.14d #25; f.17c #19)	neurocranium length from base of occipital condyle to Nasion	distance: foramen magnum to frontal midpoint	basicranial axis (f.14c #4; f.17c #4)

162 Approaches to Faunal Analysis

TABLE 2 (continued)

Position M1 (Columns 37-40)	Position M2 (Columns 41-44)	Position M3 (Columns 45-48)	Position M4 (Columns 49-52)
Skull [carnivores] (04)			
length of braincase (f.14 #11)	breadth external to auditory meatus (f.14c #24)	skull height (f.14b #38)	----
Skull [carnivores] (07)			
length of cheektooth row (f.14a #15; f.17a #12)	length of molar row (f.14c #16)	length of premolar row (f.14c #17; f.17c #13)	----
Skull [carnivores] (09)			
condylobasal length (f.14c #2; f.17c #2)	greatest length (f.14a #1; f.17a #1)	basal length (f.14c #3; f.17c #3)	----
Teeth [abridged code - carnivores] (08, 13)			
length of P2	greatest diameter of the canine (f.22b #21)	smallest diameter of the canine	length of P4 (f.15a; f.16a)
Teeth [abridged code - artiodactyls] (08, 13)			
radius of inner curve of the canine (Sus)	length of M1	breadth of M1	length of M2
Teeth [complete code] (91, 92)			
length of P1	breadth of P1	length of P2	breadth of P2
Teeth [complete code] (93, 94)			
length of M1	breadth of M1	length of M2	breadth of M2
Mandible, part 1 [herbivores] (11)			
length of cheektooth row (f.19a #6; f.20 #6; f.21a #7; f.25 #2)	length of molar row (f.19a #7; f.20 #7; f.21a #8)	length of premolar row (f.19a #8; f.21a #9)	length of diastema (f.19a #15; f.20 #9; f.21a #11; f.25 #4)
Mandible, part 1 [carnivores, Sus] (11)			
length of cheektooth row (f.22b #7; f.23a #8; f.24 #5)	length of molar row (f.22b #8; f.23a #10)	length of premolar row (f.22b #9; f.23a #11)	length P_2-P_4 (f.22b #9a; f.23a #12)
Mandible, part 2 (12)			
length from Gonion caudale to most aboral indentation of mental foramen (f.21a #6, and to be taken for all other species)	length from indentation between condyle process and angular process to aboral border of Canine (f.23a #5; f.24 #4)	length from Gonion caudale or indentation between condyle process and angular process to Infradentale (f.19a #1; f.20 #1; f.21a #1; f.22b #1; f.23a #3; f.24 #2; f.25 #1)	length from indentation between condyle process and angular process to aboral border of I_3

Position M5 (Columns 53-56)	Position M6 (Columns 57-60)	Position M7 (Columns 61-64)	Position M8 (Columns 65-68)
Skull [carnivores] (07)			
----	greatest length of nasals (f.14a #10)	basifacial axis (f.14c #5; f.17c #5)	greatest palatal breadth (f.14c #34; f.17c #26)
Skull [carnivores] (09)			
----	greatest breadth (f.14a #30; f.17a #23)	----	----
Teeth [abridged code - carnivores] (08, 13)			
breadth of P4 (f. 15a; f.16a)	length of M1 (f.15b,d; f.16b,e; f.23b)	breadth of M1 (f.15b,d; f.16b,e; f.23b)	length of P3
Teeth [abridged code - artiodactyls] (08, 13)			
breadth of M2	length of M3 (f.6b; f.19b; f.21b; f.22a)	breadth of M3 (f.6b; f.19b; f.21b; f.22a)	length of last deciduous premolar
Teeth [complete code] (91, 92)			
length of P3	breadth of P3	length of P4	breadth of P4
Teeth [complete code] (93, 94)			
length of M3	breadth of M3	great diameter of canine	small diameter of canine
Mandible, part 1 [herbivores] (11)			
length of ramus from aboral border of M_3 to aboral border of I_4	smallest height of the diastema	height of mandible in front of M_1 (f.19a #22b; f.21a #15b)	height of mandible behind M_3 (f.19a #22a; f.20 #13; f.21a #15a)
Mandible, part 1 [carnivores, <u>Sus</u>] (11)			
length from aboral border of M_3 to aboral border of Canine (f.22b #6; f.23a #7)	height in front of P_1	height in front of M_1 (f.22b #16b; f.23a #19)	height behind M_3 (f.22b #16a)
Mandible, part 2 (12)			
length from process angularis to aboral border of Canine (f. 23a #6)	length from process angularis to Infradentale (f.23a #2)	length from condyle process to aboral border of Canine (f.23a #4; f.24 #3)	length from condyle process to Infradentale (f.19a #2; f.20 #2; f.21a #2; f.22b #2; f.23a #1; f.24 #1)

TABLE 2 (continued)

Position M1 (Columns 37-40)	Position M2 (Columns 41-44)	Position M3 (Columns 45-48)	Position M4 (Columns (49-52)
Scapula [figure 31] (23)			
SLC	GLP	LG	BG
Humerus [figure 32] (24)			
Bd	BT	SD	Bp
Radius [figure 33] (25)			
Bp	SD	Bd	GL
Ulna [figure 33] (26)			
BPC	DPA	SDO	GL
Carpus [figure 40] (28)			
Breadth	Depth	Length	----
Metacarpus [figure 44] (30-34)			
Bp	Dp	SD	Bd
Pelvis [figure 34] (41)			
LA	GBA	lateral length	GL
Femur [figure 35] (42)			
Bp	DC	SD	Bd
Tibia [figure 37] (44)			
Bp	SD	Bd	Dd
Fibula [figure 38] (45)			
GL	GL without epiphysis	----	----
Metatarsus [figure 44] (50-54)			
Bp	Dp	SD	Bd
Patella [figure 36] (43)			
GL	GB	greatest depth	----
Os malleolare [figure 39] (45)			
L	GD	----	----
Talus (Astragalus) [figure 41] (46)			
GL1	GLm	Dl	Dm
Calcaneus [figure 42] (47)			
GL	GB	greatest lateral depth	smallest depth of Tuber calcis

Position M5 (Columns 53-56)	Position M6 (Columns 57-60)	Position M7 (Columns 61-64)	Position M8 (Columns 65-68)
Scapula [figure 31] (23)			
height of column	HS	DHA	Ld
Humerus [figure 32] (24)			
Dp	GLC	GL	GL without epiphysis
Radius [figure 33] (25)			
GL without distal epiphysis	GL without proximal and distal epiphyses	BFp	BFd
Ulna [figure 33] (26)			
GL without epiphysis	----	----	----
Metacarpus [figure 44] (30-34)			
Dd (not shown)	GL	GL without epiphysis	CD
Femur [figure 35] (42)			
breadth of trochlea patellaris	GLC	GL	GL without epiphyses
Tibia [figure 37] (44)			
GL	GL without proximal epiphysis	GL without proximal and distal epiphyses	Ll
Metatarsus [figure 44] (50-54)			
Dd (not shown)	GLl (Equus) GL without proximal projection (ruminants)	GL	CD
Talus (Astragalus) [figure 41] (46)			
Bd	----	----	----
Calcaneus [figure 42] (47)			
GL without epiphysis	----	----	----

TABLE 2 (continued)

Position M1 (Columns 37-40)	Position M2 (Columns 41-44)	Position M3 (Columns 45-48)	Position M4 (Columns 49-52)
Tarsus [figure 43] (49)			
GB	depth	L	----
Phalanx 1 [figure 45] (35, 55, 65)			
Bp	SD	Bd	GLpe
Phalanx 2 [figure 46] (36, 56, 66)			
Bp	SD	Bd	GL
Phalanx 3 [figure 48] (37, 57, 67)			
GL	Ld	HP	GB
Atlas [figure 27] (70)			
GB	BFcr	BFcd	GL
Axis [figure 28] (71)			
BFcr	SBV	BPacd	LCDe
Other vertebrae [figure 30] (72-74)			
PL	height of vertebral canal	----	----
Sacrum [figure 29] (75)			
GB	BFcr	length of corpus (body)	number of segments measured

| Position M5 (Columns 53-56) | Position M6 (Columns 57-60) | Position M7 (Columns 61-64) | Position M8 (Columns 65-68) |

Phalanx 1 [figure 45] (35, 55, 65)

Dp	smallest depth	depth of distal end	physiological length as illustrated by dotted lines in f.45e

Phalanx 2 [figure 46] (36, 56, 66)

Dp	smallest depth	depth of distal end	----

Phalanx 3 [figure 48] (37, 57, 67)

BF	depth of the Facies articularis	----	----

Atlas [figure 27] (70)

GLF	height of vertebral canal	----	----

Axis [figure 28] (71)

LAPa	diameter of vertebral canal	H	----

"BONECODE" - A SYSTEM OF NUMERICAL CODING FOR FAUNAL DATA FROM MIDDLE EASTERN SITES

Richard H. Meadow, Department of Anthropology
Peabody Museum, Harvard University

Introduction

BONECODE is the end product of several years of designing and testing systems for the computer processing of faunal material from archaeological sites in the Middle East. Attaining its present form in 1975, the origins of the system reach back to a paper circulated in Budapest by Charles Reed in 1971. When, during the next three years, early versions of BONESORT and KNOCOD were circulated by their developers, features of each were selected for incorporation into what came to be known as BONECODE. The author is indebted to Richard Redding, Melinda Zeder, John McArdle, and Hans-Peter Uerpmann for making available copies of their codes, both now published in this volume in more developed form.

The Code

Like other coding systems, data input for BONE-SORT is based on the 80-column punch-card. In order, however, to accommodate a full range of attribute, measurement, and contextual information for each specimen, four different input formats were designed, each designated by a different number (1-4) in column 11 of the punch-card. The arrangement, labels, and names of the data fields for each of these format types are displayed in Table 1. Cross-referencing of input is accomplished by the first ten columns of the Post-cranial (1), Cranial (2), and Measurement (3) cards and by the first six columns of the Context (4) card. Each specimen analyzed is represented on at least one Cranial or Post-cranial card. One or more Measurement cards may or may not be present. Each group of specimens which come from the same site and archaeological context share a Context card thus eliminating the need to repetitiously code archaeological information.

Table 2 lists the possible states which each of the data fields (attributes) may assume along with the output codes used in the author's print programs. Notes on various of the data fields follow at this point.

BON (Bone ID Number). Generally specimens are numbered consecutively within a locus. Each separate find and not each element gets a different number unless a specimen with many elements (e.g., a jaw with many teeth or a skull fragment with more than one "bone") is to be coded using only the "Post-cranial" format (1). Specimens which articulate or are thought to do so are given consecutive numbers if possible. Radius and ulna when fused together but coded separately get consecutive numbers with the radius receiving the lower number.

ART (Articulation Code). Formats 1 and 2: For specimens which articulate or are thought to do so, the data card for the first is coded '2' and those for the others are given the appropriate code from '3' to '7'. Otherwise a '1' is entered in the field. Format 3: The codes from '1' to '9' together with the element identification code (ELM) define the nature of the set of measurements included on a given Measurement format (3) card. The field is coded '1' for post-cranial material. A skull, however, may require more than one measurement card.

ELM (Element Identification). When employing the Cranial format (2), the code '001' is marked except when fragmented specimens bearing teeth or when horncores or antlers are being recorded in which case the appropriate code from '016' to '023' is generally used.

ANM (Animal Identification). This list is prepared for use with mammalian fauna from the Middle East. The codes from '001' to '008' are used to classify unidentifiable bone finds and were adopted from Uerpmann's KNOCOD system (this volume). The list of taxa is drawn from Ellerman and Morrison-Scott (1966) and Lay (1967) with certain modifications based on more recent work with various groups such as the gazelles and the felids (Lange 1972 and Leyhausen 1975 respectively). In the case of both ELM and ANM, additions can be made so that the data fields are useable for coding remains of birds, fish, reptiles, and amphibians.

DOM (Domestication). There are basically three ways to distinguish wild from domestic animals from the macroscopic examination of bone finds. These are: 1) identifying bones as coming from taxa normally thought of as wild (e.g., gazelle, onager, mongoose); 2) using qualitative morphological attributes such as horncore shape; and 3) plotting measurements and noting differences in size and proportions of certain specimens. This coding field is used only for the first two approaches. Otherwise it is coded '0'.

SYM (Symmetry). Phalanges of artiodactyls can generally be distinguished as being either from the medial side of the right limb or the lateral side of the left limb or from the lateral side of the right limb or the medial side of the left limb. Thus the codes '4' and '5' are included.

AGE (Relative Age). This data field is adopted from KNOCOD. See Uerpmann's discussion of it in this volume as well as Bökönyi (1970).

SIZ (Relative Size). The various states within this field are determined subjectively within each age group. See Bökönyi (1970).

MOD (Modification). If multiple conditions are present on a specimen, code the highest applicable number from the list and note the others in the comment section.

PCS (Number of Pieces). This field is for recording the number of pieces included on a data card. The code '01' is employed unless multiple fragments are described together. The code '00' is used if a separate card has already been made out for another element to which this second element is attached.

ELS (Number of Elements). The number of elements included on one data card will equal the number of pieces unless a bone find is composed of more than one element (e.g., jaw with teeth, skull, ulna fused with radius and coded together). For a jaw with six teeth the number of pieces (PCS) is '01' and the number of elements (ELS) is '07'. For fragments which cannot be identified as to element, either '--' or '00' is coded in this field.

The Post-cranial and Cranial formats (1 and 2) are identical through input column 37. Thereafter they diverge. No special notes are required for the Cranial format.

FSZ (Fragment Size). This field is used to code the best estimate of the size of the specimen at the time of discard (except states '0' and '9').

END (Proximal/Distal). The categories noted here are self-evident when they apply to long bones. For such elements as the scapula, ilium, ischium, pubis, ulna, and calcaneum the user must define the areas of the bone to which these terms refer or add to the list of states published here.

PRF, DSF, PPF, SHF, PAP, PML, DAP, DML (Fracture characterization). The data fields for Fracture Type (PRF and DSF), Position of Primary Fracture (PPF) and Position of Fracture: Shaft (SHF) are used only in describing ancient breaks on long bones. They are abstracted from the work of Hind Sadek-Kooros (1966, 1972, 1975) which should be referred to before attempting to use these categories. The term 'anterior' is used here to refer to the 'cranial' or 'dorsal' aspect of a piece while the term 'posterior' designates the 'caudal' or 'volar' or 'plantar' aspect. The fields SHF and PAP can also be used to describe more completely fragmentation of the pelvis and the scapula.

TTY-TWR (fields for loose teeth). Loose teeth are coded using the 'Post-cranial' format (1). Wear categories are the same as those used in the Cranial format (2). The wear stages defined by Payne (1973) for sheep and goat mandibles are illustrated in Figure 1 together with their corresponding input and output codes.

The Measurement format (3) data fields coincide with those of formats 1 and 2 through column 15 (ELM). Thereafter, each block of five columns is used to record a separate modifier and measurement, with up to twelve measurements per card. The qualifier codes are defined as follows:
1 measurement taken
2 measurement estimated
3 measurement influenced by pathology
4 see comment
5 end measured is not fused.

The individual dimensions to which each of the twelve measurement fields refer for any given element are not defined here. Field definitions such as those made by Uerpmann for KNOCOD (this volume) can be used.

The Context format (4) coding fields and states were defined with the assistance of Thomas W. Beale specifically for use with the archaeological collections from Middle Eastern mound or *tell* sites. They require no special comment.

TABLE 1. BONECODE input format (80-column punch-card)

POSTCRANIAL	CRANIAL		MEASUREMENT	CONTEXT
SIT=site identification number	SIT	01 02	SIT	SIT
LOC=locus identification number	LOC	03 04 05 06	LOC	LOC
BON=bone identification number	BON	07 08 09 10	BON	SMJ=square - major division SMN=square - minor division
FRM=input format code (1)	FRM=input format code (2)	11	FRM=input format code (3)	FRM=input format code (4)

POSTCRANIAL	CRANIAL	#	MEASUREMENT	CONTEXT
ART=articulation code	ART		ART=measurement group	YRE=year of excavation (last two digits)
		12		
ELM=ID as to element	ELM	13	ELM	TMJ=test trench - major
		14		TMN=test trench - minor
		15		
ANM=ID as to taxon	ANM	16	MD1=modifier #1	LMJ=stratum (level) - major division
		17		
		18	MS1=measurement #1 (###.# millimeters)	LMN=stratum - minor
DOM=domestication state	DOM	19		FMJ=feature - major division
CER=certainty of ID	CER	20		
SYM=symmetry (rt./lft.)	SYM	21	MD2	FMN=feature - minor
SEX=sex determination	SEX	22		FFF=floor/feature/fill
AGE=age stage estimate	AGE	23	MS2	
SIZ=relative size	SIZ	24		INX=interior/exterior
DIS=pathology (disease)	DIS	25		ICX=interior context
BUR=burning traces	BUR	26	MD3	ECX=exterior context
MOD=bone modification	MOD	27		FIL=fill description
FRG=fragmentation	FRG	28	MS3	
PCS=number of pieces=01 unless grouped	PCS	29		QAL=context quality
		30		RCV=recovery type
ELS=number of elements (6 teeth + 1 jaw = 7)	ELS	31	MD4	
		32		SAM=same context as (4-digit locus no.)
		33	MS4	
WGT=weight to nearest half gram (###.#)	WGT	34		
		35		
		36	MD5	CTM=contemporary context (4-digit locus no.)
COM=comment	COM	37		
FSZ=fragment size	RHC=rt.horncore(antler)	38	MS5	
END=proximal/distal	LHC=lft.horncore(antler)	39		
	FRN=frontal	40		EAR=immediately earlier context (4-digit locus no.)
FUS=state of fusion	PAR=parietal	41	MD6	
	INP=interparietal	42		
PRF=prox.fracture type	OCC=occipital	43	MS6	
DSF=dist.fracture type	BAS=basio-occipital	44		LAT=immediately later context (4-digit locus no.)
PPF=position prim.fract.	SPH=sphenoid	45		
SHF=position of fracture (shaft or dors/vent)	PTR=pterygoid	46	MD7	
	PRT=petrous	47		PR1=period of occupation #1
PAP=prox.fracture(A/P)	TEM=temporal	48	MS7	
PML=prox.fracture(M/L)	VOM=vomer	49		PH1=phase of occupation #1
DAP=dist.fracture(A/P)	PAL=palatine	50		
DML=dist.fracture(M/L)	ZYG=zygomatic	51	MD8	SB1=subphase of occupation #1
CAP=cuts on shaft A/P	LAC=lacrimal	52		
CML=cuts on shaft M/L	NAS=nasal	53	MS8	PR2=period of occupation #2 (inclusive)
CPA=cuts on prox.end A/P	MAX=maxillary	54		
CPM=cuts on prox.end M/L	PMX=premaxillary	55		PH2=phase of occupation #2 (inclusive)
CDA=cuts on dist.end A/P	AMN=mandible-anterior	56	MD9	
CDM=cuts on dist.end M/L	PMN=mandible-posterior	57		SB2=subphase of occupation #2
TTY=tooth type	CUT=cuts	58	MS9	
TCL=tooth class		59		SCH=site characteriz.
TNO=tooth number	WI1=wear of incisor #1	60		THO=thousand years
TRT=tooth root	WI2=wear of incisor #2	61	MD10	
TWR=tooth wear	WI3=wear of incisor #3	62		HUN=hundred years
	WCN=wear of canine	63	MS10	BAC=B.C./A.C.
blank	WP1=wear of premolar#1	64		CMN=comment follows
	WP2=wear of premolar#2	65		blank
comment	WP3=wear of premolar#3	66	MD11	comment
		67		
	WP4=wear of premolar#4	68	MS11	
		69		
	WM1=wear of molar #1	70		
		71		
	WM2=wear of molar #2	72	MD12	
		73	MS12	
	WM3=wear of molar #3	74		
		75		
	comment	76	comment	
		77		
		78		
		79		
		80		

Acknowledgment

Besides those persons mentioned in the text, I owe a debt of gratitude to Barbara Lawrence for giving good advice during the development of BONE-CODE.

References

Bökönyi, S.
   1970 "A new method for the determination of the number of individuals in animal bone material," *American Journal of Archaeology,* vol. 74, pp. 291-292.

Ellerman, J.R. and T.C.S. Morrison-Scott
   1966 *Checklist of Palaearctic and Indian Mammals, 1758 to 1946.* Second edition. London.

Lange, J.
   1972 "Studien an Gazellenschädeln. Ein Beitrag zur Systematik der kleineren Gazellen, *Gazella* (De Blainville, 1816), *Säugetierkundliche Mitteilungen,* vol. 20, pp. 193-249.

Lay, D.M.
   1967 *A Study of the Mammals of Iran.* Fieldiana: Zoology, vol. 54.

Leyhausen, P.
   1975 *Verhaltensstudien an Katzen.* Vierte Auflage. Berlin.

Payne, S.
   1973 "Kill-off patterns in sheep and goats: the mandibles from Aşvan Kale," *Anatolian Studies,* vol. 23, pp. 281-303.

Reed, C.A.
   1971 "New method for recording and analyzing faunal material from archaeological sites," manuscript circulated at the Section on Animal Domestication of the Third International Congress of the Museums of Agriculture at Budapest, 19-23 April, 1971.

Sadek-Kooros, H.
   1966 "Jaguar Cave: an Early Man Site in the Beaverhead Mountains of Idaho." Ph.D. dissertation, Harvard University.
   1972 "Primitive bone fracturing: a method of research," *American Antiquity,* vol. 37, pp. 369-382.
   1975 "Intentional fracturing of bone: description of criteria," in A.T. Clason, editor, *Archaeozoological Studies,* pp. 139-150. Amsterdam/New York.

TABLE 2. BONECODE input and output codes

SIT Site ID Number (input columns 1-2)
                    [all formats]
input   output

01   SITE NAME (16 letters maximum in heading
-                     of each report)
99

LOC Locus ID Number (input columns 3-6)
                     [all formats]
input   output

0001   0001
-           -
9999   9999

BON Bone ID Number (input columns 7-10)
                    [formats 1, 2, and 3]
input   output

0001   0001
-           -
-           -
9999   9999

FRM Input Format Code (input column 11)
                       [all formats]
input   output

1   1   post-cranial
2   2   cranial
3   3   measurement
4   4   locus/context

ART Articulation Code (input column 12)
                       [formats 1 and 2]
input   output

1   1   (unless below)
2   2   see succeeding specimen(s)
3   3   fused with preceding
4   4   articulated with preceding,
                        weighed together
5   5   articulated with preceding,
                        weighed separately
6   6   probably articulated
7   7   possibly articulated

ELM Element Identification (input columns 13-15)

input   output

---   N.A.   not applicable
000   ????   indeterminate
001   SKULL   skull
002   BASIO   basioccipital
003   OCCIP   occipital
004   SPHEN   sphenoid

005	PTERY	pterygoid
006	VOMER	vomer
007	PALAT	palatine
008	INPAR	interparietal
009	PARIE	parietal
010	FRONT	frontal
011	PETRS	petrous
012	TEMPR	temporal
013	ZYGOM	zygomatic
014	LACRI	lacrimal
015	NASAL	nasal
016	PMX/T	premaxillary with teeth
017	PMX	premaxillary without teeth
018	MAX/T	maxillary with teeth
019	MAX	maxillary without teeth
020	MAN/T	mandible with teeth
021	MAN	mandible without teeth
022	HORN	horn core
023	ANTLR	antler
024	HYOID	hyoid
025	G.VER	general vertebra(e)
026	ATLAS	atlas
027	AXIS	axis
028	CERV	cervical vertebra
029	THOR	thoracic vertebra
030	LUMB	lumbar vertebra
031	SACR	sacrum
032	CAUD	caudal vertebra
033	RIB	rib
034	COST	costal cartilage
035	STERN	sternebra
036	SCAP	scapula
037	CLAV	clavicle
038	HUMER	humerus
039	RAD	radius
040	ULNA	ulna
041	CARPL	carpal
042	C-RAD	radial carpal [except carnivores]
043	C-INT	intermediate carpal [except carnivores]
044	C-ULN	ulnar carpal
045	C-ACC	accessory carpal
046	C-1	first carpal [inconstant in equids. e.g. suids, canids, felids]
047	C-2	second carpal [e.g. equids, suids, canids, felids]
048	C-3	third carpal [same as C-2]
049	C-4	fourth carpal
050	C-R+I	radial + intermediate carpal [e.g. carnivores]
051	C-2+3	second + third carpal [e.g. ruminants]
052	MC 1	metacarpal I
053	MC 2	metacarpal II
054	MC 3	metacarpal III
055	MC 4	metacarpal IV
056	MC 5	metacarpal V
057	MC3+4	metacarpal III + IV [ruminants]
058	MC ?	indeterminate metacarpal
059	SESAP	anterior proximal sesamoid
060	SESAD	anterior distal sesamoid
061	SESAN	anterior sesamoid
062	PH1AN	anterior phalanx 1
063	PH2AN	anterior phalanx 2
064	PH3AN	anterior phalanx 3
065	PH?AN	anterior phalanx
066	PELV	pelvis [ilium + ischium + pubis]
067	IL/IS	ilium + ischium

ELM Element Identification [continued]

input	output	
068	IL/PB	ilium + pubis
069	IS/PB	ischium + pubis
070	ILIUM	ilium
071	ISCH	ischium
072	PUBIS	pubis
073	PENIS	penis bone [e.g. carnivores]
074	FEMUR	femur
075	PATEL	patella
076	TIBIA	tibia
077	FIBUL	fibula
078	MALL	lateral malleolus
079	TALUS	astragalus (talus)
080	CALC	calcaneus
081	TARSL	tarsal
082	T-CEN	central tarsal [e.g. equids, suids, canids, felids]
083	T-1	first tarsal [occ. sep. in equids]
084	T-2	second tarsal [occ. sep. in equids; as with T-CEN]
085	T-3	third tarsal [as with T-CEN]
086	T-4	fourth tarsal [as with T-CEN]
087	T-C+4	central + fourth tarsal [e.g. ruminants]
088	T-1+2	first + second tarsal [e.g. equids, usually]
089	T-2+3	second + third tarsal [e.g. ruminants]
090	MT 1	metatarsal I
091	MT 2	metatarsal II
092	MT 3	metatarsal III
093	MT 4	metatarsal IV
094	MT 5	metatarsal V
095	MT3+4	metatarsal III + IV [ruminants]
096	MT ?	indeterminate metatarsal
097	SESPP	posterior proximal sesamoid
098	SESPD	posterior distal sesamoid
099	SESPS	posterior sesamoid
100	PH1PS	posterior phalanx 1
101	PH2PS	posterior phalanx 2
102	PH3PS	posterior phalanx 3
103	PH?PS	posterior phalanx
104	MP3+4	metapodial III + IV
105	MP ?	indeterminate metapodial
106	SES.P	proximal sesamoid
107	SES.D	distal sesamoid
108	SES.	sesamoid
109	PH1	phalanx 1
110	PH2	phalanx 2
111	PH3	phalanx 3
112	PH?	phananx
113	TTHUP	loose tooth: upper
114	TTHLW	loose tooth: lower
115	HM/FR	humerus or femur fragment
116	LG.BN	long bone fragment
117	????	indeterminate fragment
118	TTHND	tooth fragment
119		
120		

ANM Animal Identification (input columns 16-18) [Formats 1 and 2]

input	output	
000	???	indeterminate
001	SMALL	smaller than rabbit-sized
002	SM-MD	rabbit-sized to medium dog-sized
003	MED.1	medium dog-sized to medium sheep-sized
004	MED.2	medium dog-sized to wild boar-sized
005	LG-MD	wild boar-sized to small cattle-sized
006	MD-LG	medium sheep-sized to medium cattle-sized
007	LARGE	domestic cattle/red deer-sized
008	V.LG.	larger than medium cattle-sized
009	S.ART	small artiodactyl
010	M.ART	medium artiodactyl
011	L.ART	large artiodactyl
012	S.BOV	small bovid
013	L.BOV	large bovid
014	O/C/G	Ovis/Capra/Gazella (sheep/goat/gazelle)
015	OV/CP	Ovis/Capra (sheep/goat)
016	OVIS	Ovis sp. (sheep)
017	CAPRA	Capra sp. (goat)
018		
019	GAZEL	Gazella sp. (gazelle)
020	G.DOR	Gazella dorcas (dorcas gazelle)
021	G.GAZ	Gazella gazella
022	G.SGT	Gazella subgutturosa (goitered gazelle)
023		
024		
025	BOS	Bos sp. (cattle)
026	B.TAU	Bos taurus (European cattle)
027	B.IND	Bos indicus (zebu)
028	BISON	Bison bonasus (European bison, wisent)
029	BUBAL	Bubalus bubalis (water buffalo)
030	S.CER	small cervid
031	L.CER	large cervid
032	C.CAP	Capreolus capreolus (roe deer)
033	CERV.	Cervus elaphus (red deer)
034	DAMA	Dama sp. (fallow deer)
035	D.DAM	Dama dama
036	D.MES	Dama mesopotamica (Persian fallow deer)
037	SUS	Sus scrofa (boar/pig)
038		
039	CAMEL	Camelus sp. (camel)
040	EQUID	small/medium equid
041	LG.EQ	large equid
042	E.A/H	Equus asinus/Equus hemionus
043	E.ASN	Equus asinus (donkey or ass)
044	E.HEM	Equus hemionus (wild half-ass)
045	E.CAB	Equus caballus (horse)
046		
047		
048		
049		
050	S.CAR	small carnivore
051	M.CAR	medium carnivore
052	L.CAR	large carnivore
053	VIVER	viverrid
054	HERPS	Herpestes sp. (mongoose)
055	H.AUR	Herpestes auropunctatus (small Indian mongoose)

ANM Animal Identification [continued]

input	output	
056	H.ICH	*Herpestes ichneumon* (Egyptian mongoose)
057	H.EDW	*Herpestes edwardsi* (Indian grey mongoose)
058	ICHN.	*Ichnuemia albicauda* (white-tailed mongoose)
059	GENET	*Genetta genetta* (European genet)
060	HYAEN	*Hyaena hyaena* (striped hyaena)
061	S.FEL	small felid
062	M.FEL	medium felid
063	L.FEL	large felid
064	F.CAT	*Felis catus* (domestic and wild cat)
065	F.CHA	*Felis chaus* (jungle cat)
066	F.MAR	*Felis margarita* (sand cat)
067	F.MAN	*Felis manul* (Pallas' cat)
068	CARA.	*Caracal caracal* (caracal)
069	LYNX	*Lynx lynx* (lynx)
070	P.LEO	*Panthera leo* (lion)
071	P.PAR	*Panthera pardus* (leopard)
072	P.UNC	*Panthera uncia* (snow leopard)
073	ACIN.	*Acinonyx jubatus* (cheetah)
074		
075	S.MUS	small mustelid
076	L.MUS	large mustelid
077	MARTS	*Martes foina* (stone martin)
078	VORML	*Vormela peregusna* (marbled polecat)
079	MUSTL	*Mustela nivalis* (weasel)
080	MELES	*Meles meles* (badger)
081	MELLV	*Mellivora capensis* (honey badger)
082	LUTRA	*Lutra* sp. (otter)
083		
084		
085	S.CAN	small canid
086	M.CAN	medium canid
087	L.CAN	large canid
088	CANIS	*Canis* sp.
089	C.AUR	*Canis aureus* (jackal)
090	C.FAM	*Canis familiaris* (dog)
091	C.LUP	*Canis lupus* (wolf)
092		
093		
094	FENNC	*Fennecus zerda* (fennec fox)
095	VULPS	*Vulpes* sp.
096	V.CAN	*Vulpes cana* (Blanford's fox)
097	V.RUP	*Vulpes rüppelli* (sand fox)
098	V.VUL	*Vulpes vulpes* (red fox)
099		
100	URSUS	*Ursus arctos* (brown bear)
101		
102		
103	PROCV	*Procavia capensis* (hyrax)
104		
105	ERIND	erinaceid (hedgehog)
106	ERINA	*Erinaceus europaeus* (European hedgehog)
107	HEMIE	*Hemiechinus* sp.
108	PARAE	*Paraechinus* sp.
109		
110	TALPA	*Talpa* sp. (mole)
111		
112	SORCD	soricid (shrew)
113	SOREX	*Sorex* sp.
114	CROCI	*Crocidura* sp.
115	NEOMY	*Neomys* sp. (water-shrew)
116	SUNC.	*Suncus* sp.
117		
118	CHIRO	Chiroptera (bat)
119		
120	RODEN	rodent
121	SCIUR	*Sciurus anomalus* (brown squirrel)
122	CRICD	cricetid (hamster, gerbil, vole)
123	CRICN	cricetine (hamster)
124	CALOM	*Calomyscus bailwardi* (mouse-like hamster)
125	CRICE	*Cricetulus migratorius* (grey hamster)
126	MESOC	*Mesocricetus auratus* (golden hamster)
127	MICRN	microtine (vole)
128	ARVIC	*Arvicola terrestris* (water vole)
129	ELLOB	*Ellobius fuscocapillus* (mole-vole)
130	MICRO	*Microtus* sp. (vole)
131	GERBN	gerbilline (gerbil/jird)
132	GERBL	*Gerbillus* sp. (gerbil)
133	MERIO	*Meriones* sp. (jird)
134	PSAMM	*Psammomys obesus* (fat sand rat)
135	SEKEE	*Sekeetamys calurus* (bushy-tailed jird)
136	TATER	*Tatera indica* (Indian gerbil)
137	MURID	murid (rat/mouse)
138	ACOMY	*Acomys* sp. (spiny mouse)
139	APODE	*Apodemus* sp. (field mouse)
140	ARVIC	*Arvicanthis niloticus* (Nile rat)
141	MUS	*Mus musculus* (house mouse)
142	NESOK	*Nesokia indica* (bandicoot rat)
143	RATTU	*Rattus* sp. (rat)
144	DIPOD	dipodid (jerboa)
145	ALLAC	*Allactaga* sp.
146	JACUL	*Jaculus* sp.
147	MUSCD	muscardinid (dormouse)
148	DRYOM	*Dryomys nitedula* (forest dormouse)
149	GLIS	*Glis glis* (fat dormouse)
150	ELIOM	*Eliomys melanurus* (south-west asian garden dormouse)
151	HYSTR	*Hystrix indica* (Indian crested porcupine)
152	SPALA	*Spalax leucodon* (lesser molerat)
153		
154		
155	LAGOM	Lagomorpha
156	LEPUS	*Lepus capensis* (cape hare)
157		
158		
159		
160	OCHOT	*Ochotona rufescens* (rufescent pika)
.		
.		
.		
199	HOMO	*Homo sapiens* (human)
.		
.		
.		

DOM Domestication (input column 19)
                    [formats 1 and 2]
input   output

0   ND   not determinable
1   W   wild
2   ?W   probably wild
3   D   domestic
4   ?D   probably domestic

CER Certainty (input column 20)
                [formats 1 and 2]
input   output

1   1   standard identification
2   2   preliminary identification
3   3   possible identification
4   4   probable identification
5   5   definite identification of
                    potentially questionable item

SYM Symmetry (Right/Left) (input column 21)
                            [formats 1 and 2]
input   output

-                   not applicable
0   ?   not determinable
1   R   right
2   L   left
3   M   medial
4   4   right medial/left lateral
5   5   left medial/right lateral
6   B   both left and right

SEX Sex Determination (input column 22)
                        [formats 1 and 2]
input   output

0   ?   not determinable
1   M   male
2   ?M   possibly male
3   F   female
4   ?F   possibly female
5   C   castrate
6   ?C   possibly castrate
7   MC   male or castrate
8   FC   female or castrate

AGE Relative Age (input column 23)
                   [formats 1 and 2]
input   output

-                   not applicable
0   ?   not determinable
1   Y   young (not grown up at all)
2   2   foetal or infantile
3   I   infantile
4   4   infantile or juvenile
5   J   juvenile
6   6   juvenile or sub-adult
7   S   sub-adult
8   8   sub-adult or adult
9   A   adult

SIZ Relative Size (within age groups)
              (input column 24) [formats 1 and 2]
input   output

-                   not applicable
0   ?   not determinable
1   M   medium
2   L   large
3   S   small

DIS Disease (input column 25) [formats 1 and 2]
input   output

-                   not applicable
1   1   not defined below, see comment
2   2   typical alteration for that
                    element
3   3   broken without dislocation
4   4   broken with dislocation
5   5   exostosis at articular surface
6   6   exostosis far from articular
                    surfaces
7   7   not inflamed hyperplasia
8   8   holes and grinding in
                    articular surface
9   9   holes far from articular
                    surface

BUR Burning (input column 26) [formats 1 and 2]
input   output

-                   not applicable
0   ?   indeterminate
1   1   completely calcined (white)
2   2   partly calcined
3   3   partly calcined, partly
                    carbonized
4   4   carbonized (black)
5   5   partly carbonized
6   6   burnt and partly carbonized
7   7   burnt (reddish)
8   8   partly burnt
9   9   slightly burnt

MOD Modification (input column 27)
                   [formats 1 and 2]
input   output

-                   not applicable
0   ??   possibly modified, see comment
1   R1   lightly rodent gnawed
2   R2   heavily rodent gnawed
3   C1   lightly carnivore gnawed
4   C2   heavily carnivore gnawed
5   BL   blow marks present
6   RT   retouched
7   PL   polished
8   PR   pierced
9   TL   tool (shaped and used?)

FRG   Origin of Fragmentation   (input column 28)
                                [formats 1 and 2]

input	output	
-		not applicable
0	???	not determinable
1	WHL	complete and unbroken
2	PRE	pre-depositional
3	?PR	? pre-depositional
4	PST	post-depositional but pre-modern
5	?PT	? post-depositional
6	MOD	modern
7	?MD	? modern
8	P+R	pre-depositional and minor recent
9	R+P	pre-depositional and major recent

PCS   Number of Pieces   (input columns 29-30)
                         [formats 1 and 2]

input	output	
--		not applicable (e.g. fused elements)
00	00	
-	-	
99	99	

ELS   Number of Elements   (input columns 31-32)
                           [formats 1 and 2]

input	output	
--		not applicable (e.g. fragments not identifiable as to element)
00	00	
-	-	
99	99	

WGT   Weight (to nearest half gram)
       (input columns 33-36)   [formats 1 and 2]

input	output	
----		not applicable (e.g. fused elements)
0000	000.0	
-	-	
9995	999.5	

COM   Comment Follows   (input column 37)
                        [formats 1 and 2]

input	output	
0		no comment
1	1	comment
2	2	? modification (MOD, col. 27)
3	3	both '1' and '2'
4	4	small piece of radius or ulna attached to the other
5	5	both '1' and '4'
6	6	
7	7	both '1' and '6'
8	8	
9	9	both '1' and '8'

POST-CRANIAL FORMAT (format 1) ONLY

FSZ   Fragment Size   (input column 38)

input	output	
-		not applicable
0	??	not determinable
1	EP	epiphysis
2	L1	less than 1/4 estimated length (or size)
3	L2	more than 1/4 but less than 2/4 (1/2) estimated length
4	L3	more than 1/2 but less than 3/4 estimated length
5	L4	more than 3/4 but less than 4/4 (whole) estimated length
6	WH	whole length (size) but still broken
7	UN	unbroken
8	VS	various sizes grouped together
9	MD	modern break, but fragment size greater than 1/2

END   Proximal/Distal   (input columns 39-40)

input	output	
--		not applicable
00	???	not determinable
01	P-A	proximal articulation
02	P-E	proximal end
03	PES	proximal end and shaft
04	P-S	proximal shaft
05	SHF	shaft
06	D-S	distal shaft
07	DES	distal end and shaft
08	D-E	distal end
09	D-A	distal articulation
10	WHL	complete (whole)

FUS   State of Fusion   (input columns 41-42)

input	output	
--		not applicable
00	???	not determinable
01	FUS	fused
02	LIN	epiphyseal line
03	UNF	unfused
04	F/F	proximal fused/ distal fused
05	F/L	proximal fused/ distal epiphyseal line
06	F/U	proximal fused/ distal unfused
07	L/F	proximal epiphyseal line/ distal fused
08	L/L	proximal epiphyseal line/ distal epiphyseal line
09	L/U	proximal epiphyseal line/ distal unfused
10	U/F	proximal unfused/ distal fused
11	U/L	proximal unfused/ distal epiphyseal line
12	U/U	proximal unfused/ distal unfused

178 *Approaches to Faunal Analysis*

PRF Fracture Type Proximally (input column 43)

input	output	
-		not applicable
0	?	not determinable
1	1	greenstick
2	2	transverse irregular
3	3	oblique irregular
4	4	transverse irregular
5	5	oblique regular
6	6	spiral entire irregular
7	7	spiral end irregular
8	8	spiral end regular
9	9	spiral entire regular

DSF Fracture Type Distally (input column 44)

input	output	
-		not applicable
0	?	not determinable
1	1	greenstick
2	2	transverse irregular
3	3	oblique irregular
4	4	transverse regular
5	5	oblique regular
6	6	spiral entire irregular
7	7	spiral end irregular
8	8	spiral end regular
9	9	spiral entire regular

PPF Position of Primary Fracture (input column 45)

input	output	
-		not applicable
0	?	not determinable
1	1	begins at distal diaphysis
2	2	begins less than 1/4 way up shaft from distal end
3	3	begins less than 1/2 way up shaft but more than 1/4
4	4	begins less than 3/4 way up shaft but more than 1/2
5	5	begins less than whole way up shaft but more than 3/4
6	6	begins at proximal diaphysis
7		no primary breaks (= n.a.)
8	S	sliver (less than 3/5ths of circumference present)

SHF Position of Fracture: Shaft or Dorsal/Ventral (input columns 46-47)

input	output	
--		not applicable
00	??	not determinable
01	M-	begins on medial half
02	L-	begins on lateral half
03	A-	begins on anterior half
04	P-	begins on posterior half
05	MA	begins on medial/anterior quarter
06	MP	begins on medial/posterior quarter
07	LA	begins on lateral/anterior quarter
08	LP	begins on lateral/posterior quarter
09	AP	begins on anterior and posterior halves
10	ML	begins on medial and lateral halves
11	TV	transverse fracture (often not used)
		[axial skeleton and pelvis]
20	??	not determinable (= '00')
21	D	dorsal portion present
22	C	central portion present
23	V	ventral portion present
24	WH	complete dorsoventrally

PAP Position of Fracture Proximally (anterior/posterior) (input column 48)

input	output	
-		not applicable
0	?	not determinable
1	A	anterior portion present
2	C	central portion present
3	P	posterior portion present
4	W	complete anterior-posterior

PML Postion of Fracture Proximally (medial/lateral) (input column 49)

input	output	
-		not applicable
0	?	not determinable
1	M	medial portion present
2	C	central portion present
3	L	lateral portion present
4	W	complete medial-lateral

DAP Position of Fracture Distally (anterior/posterior) (input column 50)

same as for proximal end (PAP)

DML Position of Fracture Distally (medial/lateral) (input column 51)

same as for proximal end (PML)

CAP Cuts on Shaft (or dorsally) anterior/
     posterior (input column 52)

input	output	
-		not applicable
0	?	not determinable
1	A	anterior
2	P	posterior
3	B	both anterior and posterior

CML Cuts on Shaft (or dorsally) medial/
     lateral (input column 53)

input	output	
-		not applicable
0	?	not determinable
1	M	medial
2	L	lateral
3	B	both medial and lateral

CPA Cuts on Proximal End (or centrally) anterior/
     posterior (input column 54)

input	output	
-		not applicable
0	?	not determinable
1	A1	anterior at proximal end of shaft
2	P1	posterior at proximal end of shaft
3	B1	both anterior and posterior at proximal end of shaft
4	A2	anterior on proximal articular surface
5	P2	posterior on proximal articular surface
6	B2	both anterior and posterior on proximal articular surface
7	A3	anterior on both shaft and articular surface
8	P3	posterior on both shaft and articular surface
9	B3	anterior and posterior on both shaft and articular surface

CPM Cuts on Proximal End (or centrally) medial/
     lateral (input column 55)

input	output	
-		not applicable
0	?	not determinable
1	M1	medial at proximal end of shaft
2	L1	lateral at proximal end of shaft
3	B1	both medial and lateral at proximal end of shaft
4	M2	medial on proximal articular surface
5	L2	lateral on proximal articular surface
6	B2	both medial and lateral on proximal articular surface
7	M3	medial on both shaft and proximal articular surface
8	L3	lateral on both shaft and proximal articular surface
9	B3	medial and lateral on both shaft and articular surface

CDA Cuts on Distal End (or ventrally) anterior/
     posterior (input column 56)

same as for proximal end (or centrally) [CPA]

CDM Cuts on Distal End (or ventrally) medial/
     lateral (input column 57)

same as for proximal end (or centrally) [CPM]

LOOSE TEETH (format 1)

TTY Tooth Type (input column 58)

input	output	
-		not applicable
0	??	not determinable
1	UD	upper deciduous (milk)
2	UA	upper adult (permanent)
3	LD	lower deciduous (milk)
4	LA	lower adult (permanent)
5	U?	upper tooth
6	L?	lower tooth
7	?D	deciduous (milk) tooth
8	?A	adult (permanent) tooth

TCL Tooth Class (input column 59)

input	output	
-		not applicable
0	?	not determinable
1	I	incisor
2	C	canine
3	P	pre-molar
4	M	molar
5	C	cheek tooth (P or M)

TNO Tooth Number   (input column 60)

input	output	
--		not applicable
0	?	not determinable
1	1	first
2	2	second
3	3	third
4	4	fourth
5	5	first or second
6	6	second or third
7	7	third or fourth
8	8	first, second, or third
9	9	second, third, or fourth

TRT Tooth Root   (input column 61)

input	output	
--		not applicable
0	?	not determinable
1	1	open
2	2	separate but not closed
3	3	closed
4	4	swollen

TWR Tooth Wear   (input column 62)

[see CRANIAL FORMAT (format 2) for tooth wear categories]

input	output	
--		not applicable
00	??	not determinable
-		
28		
-		
91		
-		
-		
99		

CRANIAL FORMAT (format 2) ONLY

RHC Horncore [Antler] - right side or side not determinable   (input column 38)

input	output	horncore
--		not applicable
0	?	not determinable
1	1	base
2	2	corpus
3	3	tip
4	4	base + corpus
5	5	corpus + tip
6	6	some of base missing
7	7	some of corpus missing
8	8	some of tip missing
9	9	complete

[antler]

1	1	[base to beginning of first tine]
2	2	[first tine]
3	3	[base + first tine]
4	4	[fragment of column]
5	5	[column with tines]
6	6	[crown]
7	7	[single tine or fragments]
8	8	[base, first tine, and column with tines]
9	9	[complete or nearly complete brace of antlers]

LHC Horncore [Antler] - left side   (input column 39)

same as for right side [RHC]

FRN Frontal   (input column 40)

input	output	
--		not applicable
0	?	side indeterminate
1	1	2/3 or more of right side present
2	2	2/3 or more of left side present
3	3	2/3 or more of both sides present
4	4	less than 2/3 of right side present
5	5	less than 2/3 of left side present
6	6	less than 2/3 of right side present, 2/3 or more of left side present
7	7	less than 2/3 of left side present, 2/3 or more of right side present
8	8	less than 2/3 of both right and left sides present
9	9	[for MAX and PMX, teeth only; for BAS, foramen magnum area]

The input and output of the following are the same as for frontal (FRN)

PAR Parietal (input column 41)
INP Interparietal (42)
OCC Occipital (43)
BAS Basioccipital (44)
SPH Sphenoid (45)
PTR Pterygoid (46)
PET Petrous (47)
TEM Temporal (48)
VOM Vomer (49)
PAL Palatine (50)
ZYG Zygomatic (51)
LAC Lacrimal (52)
NAS Nasal (53)
MAX Maxillary (54)
PMX Premaxillary (55)

AMN Mandible - anterior (input column 56)

input	output	
-		not applicable
0	?	fragment (or portion not determinable)
1	1	incisor area
2	2	diastema area
3	3	alveolus area
4	4	horizontal ramus area (including base)
5	5	incisor + diastema areas
6	6	incisor + diastema + alveolus
7	7	all anterior areas represented
8	8	diastema + alveolus + horizontal ramus area
9	9	alveolus + horizontal ramus

PMN Mandible - posterior (input column 57)

input	output	
-		not applicable
0	?	fragment (or portion not determinable)
1	1	heel area
2	2	ascending ramus area
3	3	area of articular process
4	4	coronoid process
5	5	heel + ascending ramus area
6	6	heel + ascending ramus + articular process areas
7	7	all posterior areas represented
8	8	ascending ramus + articular process + coronoid process
9	9	ascending ramus + articular or coronoid process areas

CUT Cuts (input columns 58-59)

input	output	
--		not applicable
		cuts on mandible
01	M1	- incisor area
02	M2	- diastema area
03	M3	- alveolus area
04	M4	- horizontal ramus area
05	M5	- heel area
06	M6	- ascending ramus area
07	M7	- articular process
08	M8	- coronoid process area
09	BC	across occipital condyles
38	RH	horn core - right side
39	LH	horn core - left side
40	FR	frontal
41	PA	parietal
42	IN	interparietal
43	OC	occipital
44	BA	basioccipital
45	SP	sphenoid
46	PT	pterygoid
47	PE	petrous
48	TE	temporal
49	VO	vomer
50	PA	palatine
51	ZY	zygomatic
52	LA	lacrimal
53	NA	nasal
54	MX	maxillary
55	PM	premaxillary

TOOTH WEAR

WI1 1st Incisor (input column 60)

input	output	
-		not present/not applicable
0	?	presence of tooth not determinable (that portion of jaw missing)
1	1	diciduous tooth lightly worn
2	2	deciduous tooth more than only lightly worn
3	3	permanent tooth bud present/milk tooth heavily worn or missing
4	P	permanent tooth present but wear not determinable
5	U	permanent tooth unworn/erupting
6	S	permanent tooth slightly worn
7	M	permanent tooth moderately worn
8	H	permanent tooth heavily worn
9	V	permanent tooth very heavily worn

The input and output of the following are the same as for 1st Incisor (WI1)

WI2 2nd Incisor (input column 61)
WI3 3rd Incisor (62)
WCN Canine (='4th incisor') (63)
WP1 1st Premolar (64)
WP2 2nd Premolar (65)

WP3 3rd Premolar (input columns 66-67)

[SEE FIGURE 1 FOR OVIS/CAPRA WEAR PATTERNS]

input	output	
--		not applicable
00	??	presence of tooth not determinable (that portion of jaw missing)
01	D?	D3 wear not determinable
02	DU	D3 unworn
03	D1	D3 slightly worn 1
04	D2	D3 slightly worn 2
05	D3	D3 slightly worn 3
06	D4	D3 moderately worn 1
07	D5	D3 moderately worn 2
08	D6	D3 moderately worn 3
09	D7	D3 heavily worn 1
10	D8	D3 heavily worn 2
11	D9	D3 heavily worn 3
12	DV	D3 very heavily worn 1
13	DX	D3 very heavily worn 2
14	CM	comment on P/D 3
15	P?	P3 wear not determinable
16	U1	P3 visible in jaw
17	U2	P3 erupting
18	U3	P3 half-erupted
19	U4	P3 erupted but unworn
20	U5	P3 coming into wear
21	S+	P3 slight wear
22	M+	P3 moderate wear
23	H+	P3 heavy wear
24	V+	P3 very heavy wear
90	D?	D3 wear not determinable
91	DS	D3 lightly worn
92	DW	D3 more than only lightly worn
93	U1	P3 visible in jaw
94	P?	P3 wear not determinable
95	U2	P3 erupting
96	S+	P3 slight wear
97	M+	P3 moderate wear
98	H+	P3 heavy wear
99	V+	P3 very heavy wear

WP4 4th Premolar (input columns 68-69)

[SEE FIGURE 1 FOR OVIS/CAPRA WEAR PATTERNS]

input	output	
--		not applicable
00	??	presence of tooth not determinable (that portion of jaw missing)
14	CM	comment on P4
15	P?	P4 wear not determinable
16	U1	P4 visible in jaw
17	U2	P4 erupting
18	U3	P4 half-erupted
19	U4	P4 erupted but unworn
20	U5	P4 coming into wear
21	S1	P4 slightly worn 1
22	S2	P4 slightly worn 2
23	S3	P4 slightly worn 3
24	S4	P4 slightly worn 4
25	M1	P4 moderately worn 1
26	M2	P4 moderately worn 2
27	H+	P4 heavily worn
28	V+	P4 very heavily worn
90	D?	D4 wear not determinable
91	DS	D4 lightly worn
92	DW	D4 more than only lightly worn
93	U1	P4 visible in jaw
94	P?	P4 wear not determinable
95	U2	P4 erupting
96	S+	P4 slight wear
97	M+	P4 moderate wear
98	H+	P4 heavy wear
99	V+	P4 very heavy wear

WM1 1st Molar (input columns 70-71)

[SEE FIGURE 1 FOR OVIS/CAPRA WEAR PATTERNS]

input	output	
--		not applicable
00	??	presence of tooth not determinable (that portion of jaw missing)
01	P?	wear not determinable
02	U0	perforation in crypt
03	U1	tooth visible in crypt
04	U2	tooth erupting
05	U3	tooth half erupted
06	U4	tooth erupted but not worn
07	U5	tooth coming into wear
08	UN	(loose) tooth unworn
09	S1	tooth slightly worn 1
10	S2	tooth slightly worn 2
11	S3	tooth slightly worn 3
12	S4	tooth slightly worn 4
13	S5	tooth slightly worn 5
14	S6	tooth slightly worn 6
15	S7	tooth slightly worn 7
16	S8	tooth slightly worn 8
17	M1	tooth moderately worn 1
18	M2	tooth moderately worn 2
19	M3	tooth moderately worn 3
20	H1	tooth heavily worn 1
21	H2	tooth heavily worn 2
22	H3	tooth heavily worn 3
23	V1	tooth very heavily worn 1
24	V2	tooth very heavily worn 2
90		not applicable for molars
91		not applicable for molars
92		not applicable for molars
93	U1	tooth visible in crypt
94	P?	wear not determinable
95	U2	tooth erupting
96	S+	slight wear
97	M+	moderate wear
98	H+	heavy wear
99	V+	very heavy wear

The input and output of the following are the same as for 1st Molar (WM1)

WM2 2nd Molar (input columns 72-73)
WM3 3rd Molar (input columns 74-75)

FIGURE 1. *Input and output codes for mandibular tooth wear of sheep and goats (based on Payne 1973).*

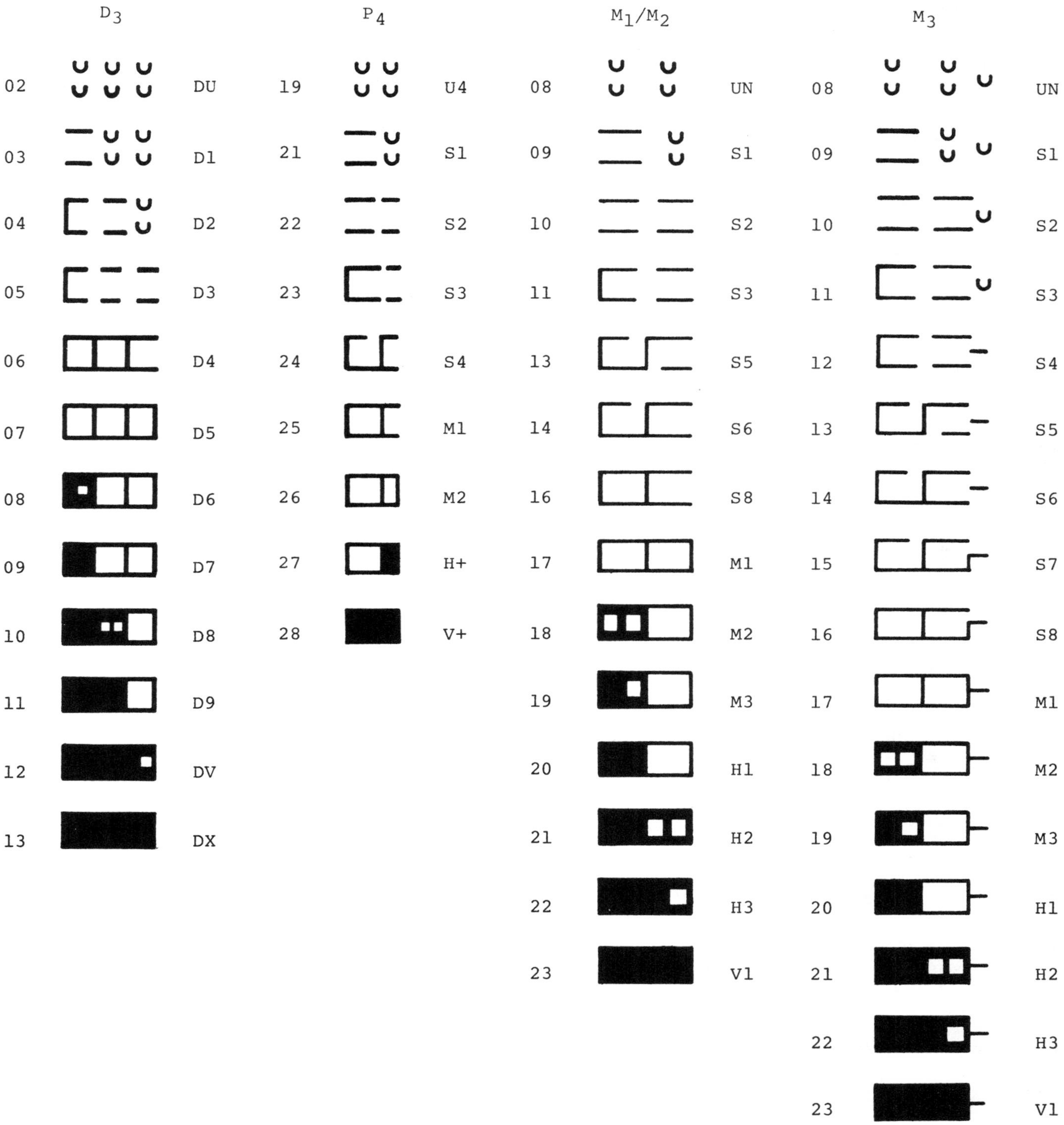

D_3		P_4		M_1/M_2		M_3	
02	DU	19	U4	08	UN	08	UN
03	D1	21	S1	09	S1	09	S1
04	D2	22	S2	10	S2	10	S2
05	D3	23	S3	11	S3	11	S3
06	D4	24	S4	13	S5	12	S4
07	D5	25	M1	14	S6	13	S5
08	D6	26	M2	16	S8	14	S6
09	D7	27	H+	17	M1	15	S7
10	D8	28	V+	18	M2	16	S8
11	D9			19	M3	17	M1
12	DV			20	H1	18	M2
13	DX			21	H2	19	M3
				22	H3	20	H1
				23	V1	21	H2
						22	H3
						23	V1

CONTEXT FORMAT

[See Table 1 for data field definitions for columns 1 through 21. Excavator's square, test trench, stratum, and feature designations are entered in this area along with year of excavation, format code (4), and the site and locus identification numbers which are used for cross-referencing.]

FFF Floor/Fill/Feature (input columns 22-23)

input	output	
--		not applicable
00	??	not determinable
01	LEVELING	leveling/intentional fill between walls
02	PLATFORM	platform
03	WALLS	walls
04	FOUNDATION	foundation trench fill
05		
06	MIDDEN	midden
07		
08		
09	MOUSE-HOLE	mouse hole
10	HEARTH	hearth
11	GRAVE	grave
12	DEPRESSION	depression
13	DRAIN	drain
14	CONTAINER	container
15	SHAPED PIT	shaped pit
16	UNSHAP.PIT	unshaped pit
17	WELL	well
18		
19		
20	MIXED FILL	fill below and above floor/surface (mixed)
21	FLR.LEVEL	approximately floor/surface level
22	ON FLOOR	on floor/surface
23	IN FLOORS	within series of floors/surfaces
24	ABOVE FLR.	immediately above floor/surface (within c. 10 cm.)
25	FLR.+ABOVE	floor/surface to floor above (may include 24 + 26)
26	ROOM FILL	fill well above floor/surface but below wall tops
27	UNDIF.FILL	fill above wall tops or undifferentiated fill
28		
29		
30	SURFACE	surface deposits

INX Interior/Exterior (input column 24)

input	output	
-		not applicable
0	??	not determinable
1	IN	inside walls
2	?IN	? inside walls
3	?OUT	? outside walls
4	OUT	outside walls
5	ASSC	associated with wall(s)
6	NONE	no significant structural remains in area

ICX Interior Context Description (input column 25)

input	output	
-		not applicable
0	??	not determinable
1	RM1	room with door
2	RM2	room with hearth
3	RM3	room without evident doorway
4	BATH	bathroom
5	SPEC	other special purpose room (note in comment)
6	OUTB	outbuilding (isolated small structure)
7		
8		
9	TOMB	tomb

ECX Exterior Context Description (input column 26)

input	output	
-		not applicable
0	??	not determinable
1	OP-F	undifferentiated open space far from buildings
2	OP-N	undifferentiated open space near buildings
3	ALLY	alley
4	STRT	street
5	CORT	court
6		
7		
8		
9		

FIL Fill Description (major ingredient) (input columns 27-28)

input	output	
--		not applicable
00	??	not determinable
01	HARD	hard or compact
02	GEL	gel (wall melt)
03	BRICK	bricky or brick debris
04	SD/BRK	sandy with brick/gel
05	SD/B/A	sandy with brick/gel and ash
06	SANDY	sandy
07	SAND/A	sandy with ash
08	ASHY	ash or ashy
09	SOFT	soft
10	SFT/BK	soft or ashy with brick/gel
11	ORGANC	organic stained, mottled, etc.
12	RUBBLE	rubble (stones with other debris)
13	WATER	water lain
14	LAMIN.	laminated
15	BURNED	burned

QAL Quality of Context (subjective assessment)
           (input column 29)

input	output	
0	?	not determinable
1	E	excellent: discrete, well-defined, single stratum/feature
2	G	good: well-defined but includes more than on stratum
3	F	fair: not well-defined but is sealed
4	P	poor: unsealed (includes unsealed pits)
5	S	surface material

RCV Recovery Type (input column 30)

input	output	
0	?	not determinable
1	WET	wet sieved
2	DRY	dry screened
3	FPK	fine picked
4	PK	picked
5	RPK	rough picked
6	OTH	combination, see comment

SAM Same Context as ... (input columns 31-34)

input	output	
----	----	not applicable
####	####	site and locus number/designation. Sometimes multiple lot numbers are used for material from the same context. If so, for one lot place its own number here and use that same number for the other relevant lots.

CTM Contemporary Context (input columns 35-38)

input	output	
----	----	not applicable
####	####	site and locus number/designation. Different context but one which has been determined through study of notes, sections, etc. to be more or less precisely contemporary.

EAR Earlier Context (input columns 39-42)

input	output	
----	----	not applicable
####	####	site and locus number/designation. The immediately earlier context with bones, in the same phase.

LAT Later Context (input columns 43-46)

input	output	
----	----	not applicable
####	####	site and locus number/designation. The immediately later context with bones, in the same phase.

PR1 Period of Occupation 1 (input columns 47-48)

input	output	
--		not determinable (see comment)
01	1	
.		
.		
30	30	
31	A	
.		
.		
56	Z	

The same codes are used for the following:

PH1 Phase of Occupation 1 (input columns 49-50)
SB1 Subphase of Occupation 1 (input columns 51-52)
PR2 Period of Occupation 2 (input columns 53-54)
PH2 Phase of Occupation 2 (input columns 55-56)
SB2 Subphase of Occupation 2 (input columns 57-58)

SCH Site Characterization (input column 59)

input	output	
0	ND	not determinable
1	CAVE	cave or rock shelter
2	CAMP	open air camp
3	STRC	isolated structure
4	DOMS	settlement: domestic area
5	CRFT	settlement: craft area
6	MONU	settlement: area of monumental architecture
7	CIRC	settlement: area of circumvallation
8	SETL	settlement: other
9	GRAV	graveyard

THO Thousand years (input columns 60-61)

input	output	
00	??	not determinable (see comment)
01	1	one thousand
.	.	etc.

HUN Hundred years (input column 62)

input	output	
0	000	even thousand(s)
.	.	
9	900	nine hundred

BAC B.C./A.C. (input column 63)

input	output	
1	BC	B.C. (Before Christ)
2	AC	A.C. (After Christ)

CMN Comment Follows (input column 64)

input	output	
0	0	no comment
1	1	comment follows
2	2	comment recorded elsewhere
3	3	both 1 and 2